One-Hour Wargames

One-Hour Wargames

By Neil Thomas

Pen & Sword
MILITARY

First published in Great Britain in 2014 and reprinted in 2025 by
PEN & SWORD MILITARY
An imprint of
Pen & Sword Books Ltd
Yorkshire – Philadelphia

Copyright © Neil Thomas, 2014

ISBN 978 1 47382 290 0

The right of Neil Thomas to be identified as
Author of this work has been asserted by him in accordance
with the Copyright, Designs and Patents Act 1988.

A CIP catalogue record for this book is
available from the British Library

All rights reserved. No part of this book may be reproduced, transmitted, downloaded, decompiled or reverse engineered in any form or by any means, electronic or mechanical including photocopying, recording or by any information storage and retrieval system, without permission from the Publisher in writing. NO AI TRAINING: Without in any way limiting the Author's and Publisher's exclusive rights under copyright, any use of this publication to 'train' generative artificial intelligence (AI) technologies to generate text is expressly prohibited. The Author and Publisher reserve all rights to license uses of this work for generative AI training and development of machine learning language models.

Typeset in Optima by Mac Style Ltd
Printed in the UK by CPI Group (UK) Ltd, Croydon, CR0 4YY

The Publisher's authorised representative in the EU for product safety is Authorised Rep Compliance Ltd., Ground Floor, 71 Lower Baggot Street, Dublin D02 P593, Ireland | www.arccompliance.com

For a complete list of Pen & Sword titles please contact

PEN & SWORD BOOKS LIMITED
47 Church Street, Barnsley, South Yorkshire, S70 2AS, England
E-mail: enquiries@pen-and-sword.co.uk
Website: www.pen-and-sword.co.uk
or
PEN AND SWORD BOOKS
1950 Lawrence Rd, Havertown, PA 19083, USA
E-mail: Uspen-and-sword@casematepublishers.com
Website: www.penandswordbooks.com

Contents

Acknowledgements		vii
Key to Symbols		viii
Chapter 1	The Practical Wargame	1
Chapter 2	Ancient Wargaming	7
Chapter 3	Ancient Wargames Rules	12
Chapter 4	Dark Age Wargaming	15
Chapter 5	Dark Age Wargames Rules	18
Chapter 6	Medieval Wargaming	21
Chapter 7	Medieval Wargames Rules	24
Chapter 8	Pike and Shot Wargaming	27
Chapter 9	Pike and Shot Wargames Rules	31
Chapter 10	Horse and Musket Wargaming	34
Chapter 11	Horse and Musket Wargames Rules	38
Chapter 12	Rifle and Sabre Wargaming	41
Chapter 13	Rifle and Sabre Wargames Rules	43
Chapter 14	American Civil War Wargaming	46
Chapter 15	American Civil War Wargames Rules	49
Chapter 16	Machine Age Wargaming	51
Chapter 17	Machine Age Wargames Rules	54
Chapter 18	Second World War Wargaming	57

Chapter 19	Second World War Wargames Rules	60
Chapter 20	Wargame Scenarios	63
Chapter 21	Wargame Campaigns	126
Chapter 22	Solo Wargaming	128
Appendix I	Background Reading	131
Appendix II	Useful Addresses	150
Index		156

Acknowledgements

I would like to thank my father, Kaye Thomas, for typing the manuscript of this book. Also to everyone at Pen & Sword Books, especially my commissioning editor, Philip Sidnell, whose enthusiasm and input have proved invaluable.

Key to Symbols

The maps in this book make use of the following symbols:

⬭	= Hill	⬬	= Lake
▨	= Wood	▦	= Marshland
▦	= Town	▨	= Deployment Zone
▬	= Road	A	= Entry Point
～	= River		
] [	= Bridge		
) (	= Ford		

Chapter 1

The Practical Wargame

A practical wargame is one that everyone can play. One would have thought that this was true of all games by definition: for if somebody can play a particular wargame, it is necessarily practicable. Feasibility is not however synonymous with accessibility, for many wargames invariably fail to account for constraints of time or space. This can be seen with reference to the wargaming ideal: a table measuring 8' x 5', and featuring hundreds of beautifully painted 28mm metal wargames figures. Such games look magnificent, and are a tribute to what can be achieved after years of effort spent amassing and painting a huge collection of wargaming material. They cannot however be described as practical for everybody: massive financial expense is required (at the time of writing, a single 28mm unpainted metal infantry figurine costs just over £1); painting hundreds of figures takes a vast amount of time; and logistics can make such games impossible. For an 8' x 5' table needs a huge amount of space – this is not a problem if the gamer has a dedicated wargames room at his or her disposal, but most players who live in average sized houses would be forced to have temporary recourse to the living room floor. This can create a whole myriad of difficulties: the game cannot be left set up overnight, for it would get in everybody else's way; other family members may well be watching television, a serious barrier to concentration; and pets can wreak havoc as they either walk over the battlefield, or play with the figures in their own somewhat idiosyncratic way!

It should be apparent that although large wargames represent the ideal to which all hobbyists should aspire, many lack the time, space or budget to make such enticing projects realizable. Yet many articles in wargames magazines assume that such enterprises are not only achievable, but commonplace – some recent pieces have for instance provided for what their writers maintain are small wargames, involving about 200 figures per side fighting over a 6' x 4' table. These do not strike me as being especially diminutive. A practical wargame on the contrary requires a genuinely small tabletop (3' x 3'); appropriately sized armies (no more than 100 figures per side); and games that can be completed in about an hour, allowing for contests in the evening after school or work.

This book is devoted to practical wargaming. It offers appropriate sets of rules, and thirty different types of battle scenarios for small tables and small armies.

The rules are simple, thereby encouraging ease of understanding and rapid play. I have included nine sets covering all the major periods. Readers should always bear in mind that simple rules are not necessarily unrealistic, which all too common misconception has resulted in some monstrously turgid and hideously complex rulebooks being produced in the name of realism. Simplicity is at least guaranteed to produce enjoyment; and I have provided introductory chapters to each of my rulesets, explaining my design parameters. This allows readers to appreciate why the rules were designed as they were, and more importantly form a critique of their efficacy. The latter activity will hopefully lead to all readers deciding to write their own rules: for designing your own wargame is a more rewarding activity than any other; you instantly become a true military historian, having researched particular conflicts and simulated them through the medium of a wargame – playing your very own military simulation will always be much more enjoyable than using someone else's rules, no matter how eminent their author may be.

The battle scenarios included after the rulesets allow for a variety of encounters, all of which can be fought using any of the rules included. They are intended to show that there are many more types of wargame than the basic competitive encounter, when two armies face each other over an essentially open battlefield, with no context provided and with the sole aim of eliminating as many enemy units as possible. Such battles can be most enjoyable, but are rather basic; more variety eventually becomes essential, and certainly provides for a more challenging (and hence rewarding) wargame.

Having a book which includes advice on producing a practical wargame is only the first step, however. The next problem lies in gathering armies and constructing a wargames battlefield. Experienced wargamers will already have both these things, and are probably advised to skip the rest of this chapter and proceed to the next. Beginners will however by definition lack both figures and a battlefield: the rest of this chapter suggests ways of acquiring both.

The first thing to do is choose a period from the nine I have provided rules for. All nine are fascinating; the reader should start collecting armies from the epoch which fires his or her imagination the most. Having done so, the next step is to acquire wargames armies. This step is greatly facilitated by visiting any large newsagent and buying copies of the latest wargaming magazines. These are always good sources of postal or website addresses of figure manufacturers, quite apart from the insights provided in the range of articles published. There is however a problem in the sheer variety of model soldiers available. The most common sizes are 28mm and 15mm, but you can also find manufacturers of 54mm, 42mm, 20mm, 10mm, 6mm and even 2mm miniatures. All these figures have their merits and their advocates; but it is fair to say that an opponent is most likely to be found if sticking to 28mm or 15mm miniatures. If further advice is required, it can always be found at a wargames show or a local club (details of both can often be found in the wargame magazines), where experienced wargamers are always ready, willing and able to help beginners into the hobby.

An alternative solution is to visit a local toy or model shop and look for the 1:72 scale plastic figures made by Airfix and other manufacturers. These have many advantages over their metal rivals. Firstly, they are extremely cheap: a box of 50 figures can be bought for about £5. They are moreover exceptionally light in terms of weight; an entire army can easily be taken to a friend's house or flat for a wargame. Thirdly, plastic figures are designed to a constant scale of 1:72 (that is to say, increasing the size of the miniature by a factor of 72 would result in a figure as tall and as big as an average human being). This is in contrast to metal figures, whose nominal 28mm size is just that: they can be anything from 27–33mm – something to consider very carefully if ordering miniatures from two different manufacturers. Plastic figures, being designed to a constant scale, are absolutely interchangeable, so that an army could consist of miniatures from a variety of companies, and still look right. Finally, 1:72 scale figures mix very well with model railway accessories, allowing the wargamer access to terrain features of exceptional quality, given that toy train enthusiasts insist upon the highest standards of aesthetic beauty in their scenery.

Having acquired some wargame figures, the next step is to paint them. This can be a very daunting prospect: the sight of beautifully painted figures in wargaming magazines can engender serious concerns that one's own efforts will always be pitifully and embarrassingly inadequate – and that it is better never to pick up a paintbrush, and give up the hobby in despair. This would be an unfortunate reaction; for alternatives to exquisite painting do exist, and are viable.

The first and almost sacrilegious option is not to use any figures at all. One can instead use pieces of card to represent the different armies. The card can simply be painted red or blue to depict the contestants, and have unit symbols marked on them. This is undoubtedly a drastic step, and would scarcely be described as aesthetically pleasing – but it does allow anyone to play a wargame very quickly, and with minimum expense.

The next alternative is almost as shocking, and this is to leave the figures unpainted. This is only possible with 1:72 plastic figures, which often feature different colours for different armies. It is, for example, common to see Second World War British infantry rendered in a khaki-coloured plastic, with their German rivals featured in an appropriate shade of grey. Similarly, American Civil War figures see Union troops often depicted in the correct shade of dark blue, and Confederates accurately rendered in grey – and if cavalry horses have chestnut coloured plastic, the results can look most effective, and an aesthetically reasonable wargame can take place.

A third option is to quite literally call in the professionals, since wargame magazines always feature advertisements from figure painters willing to paint other people's figures. The result will usually look exquisite, for the painters in question have enough of a track record to make a reasonable income from their work. These services can however be rather expensive, which can present a significant barrier to anyone on anything other than a high budget.

All the above options notwithstanding, the vast majority of wargamers will ultimately feel some degree of compulsion to paint their figures. When taking up the challenge, it is vital to consider that you only have yourself to please – you can derive just as much pleasure from a basic paint job as one involving technically sophisticated work. When producing armies for the first time, simplicity is the key, as Bruce Quarrie pointed out in his classic book, *Napoleonic Wargaming*:

> 'But do, please do, make some effort to paint them. Even if your hand isn't as steady as you would like, it isn't too much to ask of anybody a black shako (hat), red or blue jacket, and grey or white trousers, with perhaps a touch of pink for face and hands, and black shoes and musket – is it? If painted in batches of a dozen or so at a time, doing all the hats first, then all the jackets, and so on, it does not take long, and the result in terms of tabletop appearance well justifies the slight effort.'
>
> Quarrie, Bruce, *Napoleonic Wargaming* (Patrick Stephens, 1974) (p.6).

Simple paintwork such as this can be achieved quite easily. The first step is to prepare the figures. This is a simple process with metal figures, for all that is needed is to trim off any unwanted bits of surplus raw material, and attach the figures on a piece of card with a temporary adhesive such as blu-tack. They are now ready for painting. There is a slightly different approach with plastic figures, depending on the material of which they are made. When dealing with the now widely available 'hard' plastic polystyrene figures, these have first to be assembled using an appropriate glue, then mounted on bases for painting in the same way as metal figures. The approach with 'soft' plastic polythene figures of the Airfix type is rather different. These have a parting agent attached to them during the manufacturing process, designed to prevent the plastic from sticking to its mould. Unfortunately, its presence also prevents paint from sticking to the figure! Polythene figures must accordingly be scrubbed thoroughly in a bowl of water with washing up liquid, before painting. This will remove enough of the parting agent to allow paint to adhere to the miniatures. Following this, all sprue attachments should be removed from the figures, apart from the one fixing the base of each miniature. This serves to mount all figures securely, so that they do not have to be touched during the painting process.

Once the preparation has been completed, painting can occur. The first step is to undercoat several figures (anything between 8 and 24 should suffice) by providing a liberal covering of black paint and leaving overnight to dry. This undercoat allows natural shading for the miniatures, and also means that any areas missed during the painting process will not stand out too much – it simply looks like natural shadow. When painting begins, an assembly line process can be adopted. Let us assume that the budding wargamer has acquired some British Napoleonic infantry of the Waterloo campaign, and has prepared 12 figures ready for painting by giving them a black undercoat. The next step is to paint all the jackets in red. Once the last figure has had its jacket painted, the first miniature should be dry enough to paint the trousers in grey. This is followed by the hands and face (flesh or pink paint), and finally the base (green). The black undercoat can legitimately suffice for the other parts of the figure, covering muskets, shoes, backpack and headwear. Variants on this theme will cover all periods; for example, cavalry horses can be painted chestnut brown, and armour rendered in silver.

Following the painting process, metal and hard plastic figures can be mounted on their bases (card or plastic painted green) ready for action. Soft plastic figures do however need a coat of varnish, to prevent the paint from falling off. This process is achieved using a mixture of non-toxic PVA glue and water – and while on the subject of toxicity, it is definitely best to use water-based acrylic paints on figures; the enamel alternative is most unsatisfactory, relying as it does on some rather toxic liquids such as turpentine to thin the paint. Acrylic paints are thinned using water, which is clearly a much safer option all round.

The result of this painting process may appear crude, but is functional: armies can be painted and ready for tabletop action very quickly – and having completed two armies you can always acquire more. By this stage, it is possible that you may wish to experiment with more advanced painting techniques; advice on these can be found in wargaming magazines and commercially produced painting guides. Those who wish to buy hard plastic figures from Games Workshop stores (the Bretonnia and Empire ranges provide suitable miniatures for the Medieval or Pike and Shot periods), will find enthusiastic and knowledgeable staff eager to give advice on painting techniques too.

Having acquired a pair of wargame armies, the final step is to obtain some terrain for the troops to fight over. There are two principal tabletop options. The first is to drape a green cloth over a suitable table; the second is to buy an appropriate piece (3' x 3') of hardboard or chipboard from a suitable hardware store, and paint it with the requisite shade of green emulsion. The next step is to fashion the necessary terrain features, which in the context of this book means hills, rivers, lakes, roads, woods, marshes and towns. By far the best approach is to buy the scenery you need from retail outlets such as model railway suppliers, model shops, or Games Workshop stores. Those without a readily accessible retail outlet can consult wargames or model railway magazines to find suitable suppliers. Commercially available terrain will always look good, and also involve minimal effort on the part of the purchaser.

It is however possible to make your own terrain, and to do so very cheaply. Hills can be fashioned simply by draping a green cloth over a pile of books, for example. Alternatives can take the form of placing several layers of card on top of each other, and painting the top layer green; or buying some thick polystyrene from an appropriate DIY or hardware store, cutting it to the requisite shape, and painting it to the desired colour. Rivers can take the form of an old pair of jeans cut as desired; or by painting card the correct shade of blue (an approach that also works rather well with lakes). Indeed, card can be pressed into service with most types of terrain: roads can use unpainted strips; towns can be produced from buildings made of card and painted as desired; and marshes can have a card base painted green, and having tufts of carpeting of an appropriate light brown shade (usually obtainable from suitable retailers as samples) blu-tacked on top. So far as woods are concerned, I would strongly advise anyone to buy readymade trees; it is however possible to produce some basic forestry by using twigs or cocktail sticks attached to a plasticine or blu-tack base (painted green) and attaching sponges (available from a chemist) on the end of the twig or toothpick and paint them green.

Having acquired both figures and terrain, the next step for the practical wargamer is to have a set of playable rules. The next eighteen chapters cover all nine major periods – readers are encouraged to find the one that suits them, and give it a try!

Chapter 2

Ancient Wargaming

Any budding Ancient wargame designer instantly hits a snag when contemplating the period in question – specifically, just how wide a time span should his or her rules cover. Most rulesets do for example attempt to encapsulate warfare from 3000 BC to AD 1500; whilst feasible, the resultant rulebooks tend to have broad generalizations, somewhat unwieldy mechanics, and occasionally extreme complexity. I have accordingly focussed upon a shorter time span for the ruleset printed in the next chapter, allowing as it does for every simple and hence accessible gaming mechanics. My chosen period is 500 BC to AD 100, which encapsulates the zenith of the Classical Age. It covers both Greece and Rome, and includes such inspiring historical figures as Alexander the Great, Hannibal, and Julius Caesar.

The Classical period was noted for seeing the decline and fall of a purely aristocratic culture, both in political and military terms. For our epoch began with the dominance of the Achaemenid Persian empire, whose governance and operation was focussed upon a military aristocracy. This manned the ranks of the cavalry, which would deal with its rivalry by indulging in skirmishing with javelins at point blank range, and duelling with individual opponents using swords. Cavalry charges on the medieval model were rare, due primarily to the lack of stirrups and hence limited stability in the colossal impact resulting when the charge met the enemy. The Persian horsemen were ably supported by archers, whose volleys of arrows would easily account for the ill-equipped levies forming the ranks of the Persians' Asiatic enemies.

The Achaemenid imperial model was however challenged by the Greeks, whose military culture developed according to the very different political structure of the city state. These small entities spent a lot of their time fighting each other, which led to a passionate belief in the virtues of independence. This in turn saw a successful challenge to the dominance of a narrow aristocratic elite, following the rise of a substantial group of prosperous farmers. The latter could not afford to equip themselves as horsemen, but instead developed the concept of the citizen army by becoming heavy infantrymen, or hoplites. They were well armoured, carried a large shield or hoplon (hence the derivation of 'hoplite'), and a long thrusting spear. They created formidable units of deep formations, which became known as phalanxes; other troops such as skirmishers (equipped with javelins) and cavalry constituted a very small part of any Greek army, usually being assigned the task of protecting its flanks.

The Greek army provided the model for all successful classical rivals. The precise significance of the subordinate arms differed in each however, as did the exact variety of heavy infantry. Thus it was that barbarian Celtic armies specialized in foot soldiery which operated in loose formations which moved rapidly, and could launch a ferocious charge; they did however lack armoured protection, rendering them vulnerable in a sustained clash. Roman armies conversely operated in close order formation and were heavily armoured; but their equipment of a heavy short-ranged javelin and a short thrusting sword known as the gladius, made their army much more flexible than the Greek phalanx.

All this can be reflected in any simple wargame, and I have done so by allowing four major troop types to play a role on the wargames table.

1. INFANTRY

This class covers all heavy infantrymen, and always makes up at least fifty per cent of any wargames army. They operate in close order formation, resulting in them being rather slow. They are however very well armoured and shielded, and extremely potent in hand-to-hand combat.

2. ARCHERS

These troops are based upon the Persian model. They move in close order at the speed of Infantry, but rely upon their bows for impact on the battlefield. This makes them effective at a distance, but quite vulnerable in hand-to-hand combat, especially given their lack of heavy armour.

3. SKIRMISHERS

These men operated in open order and were unarmoured. This allows them to move rather more quickly than Infantry or Archers, and also enables them to operate effectively in dense terrain such as woods. They relied upon the nuisance value of skirmishing at a distance with javelins; their lack of protection and dispersal formation renders them very vulnerable in hand-to-hand combat.

4. CAVALRY

Mounted troops moved rapidly, allowing them to outflank more sedentary foes. They were quite well armoured, but the lack of protection for their horses means that Cavalry are effectively only as durable as Archers in the wargame. Horsemen relied upon skirmishing at point blank range with javelins, and individual duels with swords as mentioned earlier. This means that they are about as effective as Archers in hand-to-hand combat, but lack any long range missile capability.

This reduction of combat categories to just four types inevitably rules out some troop varieties such as horse archers, scythed chariots and elephants; it does however give some approximation of ancient battlefield activity, and allows for interesting challenges in the coordination of disparate troop types. I have simplified the depiction of units by keeping precise stipulations to a minimum; wargamers simply have to deploy each unit on a frontage of 4–6 inches. In particular, there is absolutely no prescription of how many figures should constitute a given unit – the wargamer should simply rely upon what looks right, according to the size of the figures in his or her collection. This serves to avoid pedantic and unnecessary edicts concerning unit frontage, and precisely how many figures should be crammed onto each base.

The game relies upon the use of alternate turns, with one player moving, shooting and engaging in close combat, followed by the second player. This is far more manageable than the option of having both players act simultaneously, and is somewhat surprisingly more realistic. For it is only superficially true that armies in historical battles acted simultaneously: what generally happened was that one side would act, and its opponent react – and this process can be reflected quite accurately with alternate turns.

Movement is depicted according to a simple model, whereby rapidity is reflected by faster movement rates rather than, for example, allowing some units to turn more rapidly than others. Turning is instead depicted in a simple manner, by pivoting units on their central point. This avoids the complexity of wheeling manoeuvres, where wargamers have to precisely measure the movement distance of a unit's outer corner. The difficulties of turning are instead provided for by only allowing evolutions at the start and/or the end of a unit's move, but not during it. This reproduces the historical effects, but makes the tabletop process much easier. The effects of terrain are also dealt with in a straightforward manner; so that only certain types of unit may enter a particular type of difficult terrain, but that these do not have their movement restricted after entry. This avoids the unfortunate situation of (for example) allowing all units to enter woods, but giving each different type specific movement penalties – a result that arouses all kinds of confusion in the heat of a wargames battle. My rules instead only allow skirmishers to enter woods, and not suffer any movement penalty in so doing. This is much easier to remember than the convoluted and distinctly unrealistic alternative – no sensible commander would ever have contemplated sending a hoplite phalanx into a wood, which is why I don't allow any wargamer to do anything so daft either.

Most of the effects of terrain are predictable, but two do require some explanation. The lack of movement restrictions upon units entering towns, and the limited field of fire enjoyed by their occupants, appear particularly strange, for example. This is however due to the fact that 'towns' are in reality no more than tiny hamlets (conurbations seldom played any role in the average ancient battle), which can provide a degree of cover to units in the vicinity, but neither hinder movement nor offer the 360° field of fire that a more substantial strongpoint would provide. The effects of roads also require some explanation. These were usually dirt tracks, and were only usable if the unit operated in a marching column. This formation was scarcely suitable for entering combat; which is why units only enjoy a movement bonus for road travel, if they are not charging the enemy.

Skirmishers were noted for moving quite rapidly, and may also take advantage of their dispersed formation in order to pass through other units of all types – this is not something that close order units could achieve, which is why such interpenetration is only possible for Skirmishers. What may appear surprising is that Skirmishers are not permitted to combine movement with shooting – especially since they specialized in approaching the enemy, discharging their javelins, and then retiring to their original position. I have covered this in a slightly different but simpler way, by preventing moving and shooting, but by increasing the firing range of the Skirmishers' javelins to equal that of Archers' bows: the process may appear odd; the effect is accurate.

My combat rules work on the principle of having units acquire hits throughout the game, to be eliminated after garnering 15. They retain their full fighting ability until destroyed; this reflects a model whereby real casualties are at a fairly low level, but that the sustained experience of combat will steadily degrade a unit's morale, at which point it routs. This is both simple and historically accurate: most casualties in any ancient battle (and those of most other periods too) were inflicted when the enemy fled, rather than the initial clash of arms. Essentially, loss of morale is reflected in elimination, rather than having to make frequent checks on a unit's status, which tends to be a feature of complex wargames rules.

Hits are inflicted by having the attacking unit roll a die: the increased competence of Infantry is reflected by allowing them to add 2 to their combat scores, whereas the limited performance of Skirmishers is depicted by a die roll reduction of 2. Casualties can be reduced if the defending unit enjoys a terrain advantage, be that in the form of cover; occupying higher ground; or defending a river crossing (the latter two contingencies only apply in hand-to-hand combat: standing on a hilltop has little effect if being shot at!). Defending Infantry units enjoy protection conferred by armour, which allows them to suffer casualties at a reduced rate too.

Players should note that hand-to-hand combat is one sided: you only inflict casualties in your own turn. This may appear strange, given that real life mêlées were simultaneous. I resorted to it because of simplicity; it prevents players losing track of turns, but also effectively rewards taking the initiative by charging the enemy – it will allow you to strike the first blow, and gain an advantage in so doing. Rewarding positive play is always a good thing, and serves to prevent inert tactics, which in turn avoids a tedious wargame.

Hand-to-hand combat is always a fight to the finish; rapid units can choose the time and place of engagement, but should always be very careful only to commit themselves when sure of an advantage. This can best be achieved by manoeuvring around the flank of an enemy unit, which results in doubling the number of hits inflicted on the victim. However, if a unit of Skirmishers is careless enough to get caught in a frontal engagement with an enemy Infantry unit, it is guaranteed to suffer. This is absolutely as it should be; carelessness was always fatal on the real battlefield, and the consequences must be equally sanguinary on its wargaming equivalent!

Chapter 3

Ancient Wargames Rules

UNIT TYPES

This game features the unit types of Infantry, Archers, Skirmishers, and Cavalry, each of which occupies a frontage of between 4 and 6 inches. Any size or scale of figure may be used; wargamers should decide for themselves how many figures constitute a given unit.

SEQUENCE OF PLAY

Each complete turn comprises two player turns. Each wargamer follows the sequence listed below in his or her player turn:

1. Movement
2. Shooting
3. Hand-to-Hand Combat
4. Eliminating Units

1. MOVEMENT

Movement Allowances. Units may move up to the distances listed below during their turn:

Unit Type	Movement Distance
Infantry and Archers	6"
Skirmishers	9"
Cavalry	12"

Turning. Units turn by pivoting on their central point. They may do so at the start and/or the end of their move.

Terrain. Units are affected by terrain as follows:

i. **Woods**. Only Skirmishers may enter.
ii. **Towns**. These do not restrict the movement of any unit.
iii. **Marshland and Lakes**. These are impassable to all units.
iv. **Rivers**. These may only be crossed via bridges or fords.
v. **Roads**. Units moving by road increase their movement distance by 3" if their entire move is spent on the road. This bonus may not be received if charging.

Moving and Shooting. Units may not shoot if they have moved during the same turn.

Interpenetration. Only Skirmishers may pass through other units (and vice versa).

Charge Moves. Charges are resolved by moving the attacking unit into contact with its target. They are subject to the following restrictions:

i. **Turning**. A charging unit may turn once, at the start of its move. This evolution may not exceed 45°.
ii. **Limited Engagement**. Only one attacking unit may contact each face of the target (these being Front, Left Flank, Right Flank, and Rear).
iii. **Fighting**. Combat is resolved during the Hand-to-Hand Combat phase.

2. SHOOTING

Only Archers and Skirmishers may shoot, the procedure for which is as follows:

Adjudge Field of Fire. Units may only shoot at a single target within 45° of their frontal facing.

Measure Range. Archers and Skirmishers have a range of 12".

Assess Casualties. Units roll a die when shooting. Archers use the unmodified score; the result for Skirmishers is reduced by 2. The score gives the number of hits the target acquires, which is modified by the following:

i. **Cover**. Units in woods or towns only suffer half the registered number of hits (any fractions are rounded in favour of the unit shooting).
ii. **Armour**. Infantry units acquire half the number of mandated hits (any fractions should be rounded in favour of the unit shooting).

3. HAND-TO-HAND COMBAT

The procedure for Hand-to-Hand Combat is as follows:

One Sided Combat. Units only inflict casualties during their own player turn.

Assess Casualties. Units roll a die. Cavalry and Archers use the unmodified score; Infantry add 2 to the result; and Skirmishers subtract 2. The result gives the number of hits the target acquires, which is modified as follows:

i. **Terrain Advantage**. Defenders in woods, towns, on a hill, or holding a river crossing, only suffer half the indicated number of hits (fractions are rounded in favour of the attacking unit).
ii. **Armour**. Infantry units acquire half the number of mandated hits (rounding any fractions in favour of the attacking unit).
iii. **Flank or Rear Attacks**. Units engaging the enemy flank or rear inflict double the registered number of casualties.

Movement Within Combat. Hand-to-Hand Combat only concludes with the elimination of one of the contesting sides. Units may however turn to face an attack upon their flank or rear, but only if they are not simultaneously being engaged frontally.

4. ELIMINATING UNITS

Units are eliminated upon the acquisition of 15 hits.

Chapter 4

Dark Age Wargaming

Wargame designers who focus upon the Middle Ages have similar problems to those confronting Ancient period gamers – specifically the huge breadth of an epoch lasting from the fall of Rome in the fifth century AD to the rise of firearms in the fifteenth. Breaking the Medieval period into two sub-sections, each covering the more interesting aspects of the epoch, is the best way to proceed. The first of these is the Dark Ages, covering Western Europe from 600 to 1000.

The Dark Ages was defined by the fall of the Western Roman Empire. This was caused by irresistible pressure from barbarian tribes, which lacked political and military sophistication, but whose huge numbers were enough to overwhelm the Roman Empire – for, as the Soviet dictator Joseph Stalin once put it, 'quantity has a quality all of its own'.

The new barbarian kingdoms suffered something of an inferiority complex in relation to the civilization they had conquered; a culture relying upon tribal solidarity and a heroic culture based around a chieftain's warband, certainly could not compete with the intellectual and military glories of the Roman Empire. The new regimes did however embrace Christianity; and the Roman Catholic Church provided both a link with the past, and a source of much-needed unity. For Christianity provided some essential social and political bonding agents: it insisted that the poor should respect their lords and masters; and also that the ruling nobility should feel compassion for, and more importantly protect, the poor.

These twin obligations fitted the existing economic and political conditions rather well. The absence of the Roman monetary system meant that power was based upon the control of land by groups of warlords and their retinues; the most powerful of these men became kings, who controlled their realms on the basis of interlocking obligations: the poor respected their masters, and the nobles protected the poor. All had a duty to worship the Christian God. This system of interlocking social obligations on the basis of real estate, became known as feudalism.

The Western European military systems had two different approaches, both of which were based around the concept of the noble's retinue. The Frankish kingdoms in what is now France and Germany developed cavalry forces, but the English Anglo-Saxons relied upon a nobility that rode to the battlefield on horseback, but fought on foot once they got there. I have chosen to focus upon battles in Britain, the kingdoms of which largely relied upon groups of infantry equipped with long spears and shields, and operating in close formation. These so-called 'shieldwalls' were not especially well drilled, but had tremendous endurance.

The major threat to the Saxon kingdoms (and those of the Picts and Scots who fought in a similar style), came from Scandinavian Viking invaders, who tended to rely upon infantry fighting in slightly looser formations than their foes. These moved more rapidly than their Saxon opponents, and their impetuous charges had a tremendous shock effect.

My wargames rules for the Dark Ages can use the basic principles of the Ancient rules described in the preceding two chapters. There are however some differences in the troop types selected, and in particular how they operate, which are stated below:

1. INFANTRY

This category covers all foot units equipped with long spears and shields. They are assumed to operate in close order formation, and bear some resemblance to their Ancient predecessors. They do for example have a tabletop movement allowance of 6", and the tightly packed shieldwall is assumed to give similar protection to armour in Classical times. The Dark Age infantryman's lack of training is simulated by making his combat strikepower rather less than his Ancient forebear.

2. WARBAND

This class covers impetuous troops such as Vikings. Their loose order is depicted by allowing them a movement allowance of 9", and the power of their impetuous charges reflected by allowing them to enjoy a bonus on all combat dice rolls. They do however enjoy rather less protection than the stolid shieldwall infantry, and therefore suffer casualties at the normal rate of unarmoured troops.

Dark Age Wargaming   17

3. SKIRMISHERS

Some troops, usually adolescents or peasants, were equipped with javelins and ordered to skirmish at a distance, avoiding hand-to-hand combat wherever possible. They behaved exactly like their Ancient equivalents, and are treated accordingly.

4. CAVALRY

The extent to which armies in Britain used cavalry in the Dark Ages has given rise to much animated historical debate, and will doubtless continue to do so. It is generally surmised that armies of the lowland indigenous Britons used a good many units of horsemen, which reduced markedly as the Saxons pushed them back into Wales. The Picts and Scots are believed to have had some mounted warriors, whereas it is generally surmised that the Saxons and Vikings only used horses to convey their nobles to the battlefield, rather than fight upon it. I have assumed that cavalry units were present, but that they did not perform especially effectively. They can as a result be treated as their ancient predecessors.

The armies created by these rules reflect British military activity, whereby a core of Infantry units are supplemented by Warband (reflecting a contingent of Viking mercenaries), Skirmisher and Cavalry units. Readers with a desire to depict Viking invaders, as opposed to those who had settled over a long period, could easily change the army compositions (outlined in Chapter 20) by swapping the Infantry units generated on the troop selection table with Warbands; Frankish armies can similarly be depicted by swapping Infantry units with Cavalry. It could also be argued that the horsemen of the Emperors Charlemagne and Otto deserve a combat bonus to reflect their shock impact: they could for example add 2 to all combat die rolls, whereas Frankish Warbands can be assumed to be ill-trained rabble who lose their extra attacking power. Wargamers are as always encouraged to depict their favourite troop types in an appropriate way.

Chapter 5

Dark Age Wargames Rules

UNIT TYPES
This game features the unit types of Infantry, Warband, Skirmishers, and Cavalry, each of which occupies a frontage of between 4 and 6 inches. Any size or scale of figure may be used; wargamers should decide for themselves how many figures constitute a given unit.

SEQUENCE OF PLAY
Each complete turn comprises two player turns. Each wargamer follows the sequence listed below in his or her player turn:

1. Movement
2. Shooting
3. Hand-to-Hand Combat
4. Eliminating Units

1. MOVEMENT

Movement Allowances. Units may move up to the distances listed below during their turn:

Unit Type	Movement Distance
Infantry	6"
Skirmishers and Warband	9"
Cavalry	12"

Turning. Units turn by pivoting on their central point. They may do so at the start and/or the end of their move.

Terrain. Units are affected by terrain as follows:

   i. **Woods**. Only Skirmishers may enter.
   ii. **Towns**. These do not restrict the movement of any unit.
   iii. **Marshland and Lakes**. These are impassable to all units.
   iv. **Rivers**. These may only be crossed via bridges or fords.
   v. **Roads**. Units moving by road increase their movement distance by 3" if their entire move is spent on the road. This bonus may not be received if charging.

Moving and Shooting. Units may not shoot if they have moved during the same turn.

Interpenetration. Only Skirmishers may pass through other units (and vice versa).

Charge Moves. Charges are resolved by moving the attacking unit into contact with its target. They are subject to the following restrictions:

   i. **Turning**. A charging unit may turn once, at the start of its move. This evolution may not exceed 45°.
   ii. **Limited Engagement**. Only one attacking unit may contact each face of the target (these being Front, Left Flank, Right Flank, and Rear).
   iii. **Fighting**. Combat is resolved during the Hand-to-Hand Combat phase.

2. SHOOTING

Only Skirmishers may shoot, the procedure for which is as follows:

Adjudge Field of Fire. Units may only shoot at a single target within 45° of their frontal facing.

Measure Range. Archers and Skirmishers have a range of 12".

Assess Casualties. Skirmishers roll a die when shooting, and reduce the result by 2. The score gives the number of hits the target acquires, which is modified by the following:

i. **Cover**. Units in woods or towns only suffer half the registered number of hits (any fractions are rounded in favour of the unit shooting).
ii. **Shieldwall**. Infantry units acquire half the number of mandated hits (any fractions should be rounded in favour of the unit shooting).

3. HAND-TO-HAND COMBAT

The procedure for Hand- to-Hand Combat is as follows:

One Sided Combat. Units only inflict casualties during their own player turn.

Assess Casualties. Units roll a die. Cavalry and Infantry use the unmodified score; Warbands add 2 to the result; and Skirmishers subtract 2. The result gives the number of hits the target acquires, which is modified as follows:

i. **Terrain Advantage**. Defenders in woods, towns, on a hill, or holding a river crossing, only suffer half the indicated number of hits (fractions are rounded in favour of the attacking unit).
ii. **Shieldwall**. Infantry units acquire half the number of mandated hits (rounding any fractions in favour of the attacking unit).
iii. **Flank or Rear Attacks**. Units engaging the enemy flank or rear inflict double the registered number of casualties.

Movement Within Combat. Hand-to-Hand Combat only concludes with the elimination of one of the contesting sides. Units may however turn to face an attack upon their flank or rear, but only if they are not simultaneously being engaged frontally.

4. ELIMINATING UNITS

Units are eliminated upon the acquisition of 15 hits.

Chapter 6

Medieval Wargaming

The Medieval period covers the zenith of feudalism from 1100 to 1300, and more specifically the age of chivalry. The latter concept stemmed from the increasing status and battlefield dominance of the mounted nobility, facilitated as it was by the development of horsemen who were both heavily armoured and equipped with lances – the shock impact of a Medieval cavalry charge, greatly enhanced by the development of stirrups (which appeared during the eighth century), was frequently decisive on the battlefield.

The dominance of the mounted nobility was such that their position was enshrined in society. The Medieval world became divided into three estates, each of which had a vital function. Thus it was that the clergy formed the first estate, which provided for society's spiritual and cultural needs; the knights and their armed retainers formed the second estate, which upheld justice; and the labourers formed the third estate, which provided economic support for the whole. The knights therefore had every reason to see themselves as the rightful secular rulers, since they protected Christian civilization: this self-perception was even supported by literary works of the time, such as the Arthurian legends and the tales of Emperor Charlemagne and his paladins.

With societal values and literary works supporting his position, the role of the knight became almost ritualized, as was demonstrated by the growth of the tournament. This opulent display of jousting became immensely popular from the twelfth century onwards, and the most successful participants even had their achievements celebrated in biographical works. The combination of genuine martial achievements, literary glorification, and the spectacle of the tournament duel led to the development of a code of chivalry, and a general belief that the mounted knight's position in warfare and society alike was unassailable.

There was however a realization that infantry had its uses on the battlefield. They could for example provide a rallying point behind which cavalry could recover after a succession of exhausting charges; the foot soldiery could also stand and hold strategic terrain such as prominent hills. Unfortunately, the pre-eminence of the knight meant that infantry tended to be undrilled and rather unwilling levies, who were hastily equipped with spears and shields and instructed to stand firm on the battlefield – an injunction that was not always followed; for such troops may have had reasonable striking power, but could scarcely be expected to possess the *esprit de corps* that would have guaranteed endurance at the time of ultimate trial.

Two crucial developments enhanced the combat power of infantry. The first occurred when the knights themselves dismounted and fought on foot. The nobility doubtless saw this as rather humiliating, but the combination of heavy armour and martial pride improved cohesion immensely, and turned the dismounted knight into a doughty and reliable component of the infantry contingent. The second vital development was the rise of the crossbow, a weapon that had been around for some time, but which was perfected during the Middle Ages. Archers were pretty feeble in hand-to-hand combat, given the lack of spears, shield and anything other than light armour, but the mechanical power of the crossbow meant that enemy troops were now very vulnerable to missilery – only dismounted knights had enough protection to withstand the power of a crossbow bolt. The nobility were so appalled by the threat now posed by mere commoners that they even tried to implement a legal sanction against the new weapon. Thus it was that the Second Lateran Conference of 1139 banned the use of the crossbow against Christian foes (Muslims, being infidels, were not so fortunate). This injunction met with the same level of success as most legal attempts to ban the use of particular weapons – it was soon breached, and became no more than a rather meaningless curiosity.

The rules that follow are based upon the Ancient wargaming set in their general principles, but use four different troop types, which are considered below.

1. KNIGHTS

Mounted chivalry are the most numerous type of unit featured here, reflecting their dominance in the Medieval battlefield. They move as rapidly as cavalry units, reflecting their impetuosity; the latter trait also explains their great potency in hand-to-hand combat, enjoying as they do an addition of 2 to every die roll. Their rash behaviour did however induce a lack of cohesion, which is why knights do not enjoy any benefits for their armour protection – also explained by the fact that the horses were not protected as heavily as their riders.

2. ARCHERS

These are assumed to operate in close order and be equipped with crossbows. They accordingly have a minimal capacity for hand-to-hand combat, lacking as they do an effective mêlée weapon or much in the way of armoured protection – resulting in a reduction of 2 to all hand-to-hand combat dice rolls. The crossbow is conversely an exceptionally effective missile weapon, which is why dice rolls are increased by 2 when the archers are firing.

3. MEN-AT-ARMS

These are heavily armoured dismounted knights equipped with spears and shields. Their charges lack the shock impact enjoyed by noble cavalry, which is why they do not enjoy a combat bonus. Their armour does however confer a good deal of protection and hence endurance, which is why they only suffer casualties at half the normal rate.

4. LEVIES

These rather unenthusiastic infantry are equipped with spears and shields, but little in the way of armour. They strike as effectively as units of Men-at-Arms, but lack the protection bonus received by the latter.

A glance at the rules will reveal that Medieval troops can be rather inflexible. They do for instance lack any capacity for entering woods, and may never pass through each other. This is intended to reflect the undisciplined and untrained nature of many units. This does not in any way diminish the fascination of this period; it does instead provide a great test of any wargamer's ability – and those with an inclination to develop their figure painting skills, should note that noble heraldry can look very striking when rendered by gifted brushwork.

Chapter 7

Medieval Wargames Rules

UNIT TYPES

This game features the unit types of Knights, Archers, Men-at-Arms, and Levies, each of which occupies a frontage of between 4 and 6 inches. Any size or scale of figure may be used; wargamers should decide for themselves how many figures constitute a given unit.

SEQUENCE OF PLAY

Each complete turn comprises two player turns. Each wargamer follows the sequence listed below in his or her player turn:

1. Movement
2. Shooting
3. Hand-to-Hand Combat
4. Eliminating Units

1. MOVEMENT

Movement Allowances. Units may move up to the distances listed below during their turn:

Unit Type	Movement Distance
Men-at-Arms, Levies and Archers	6"
Knights	12"

Turning. Units turn by pivoting on their central point. They may do so at the start and/or the end of their move.

Medieval Wargames Rules 25

Terrain. Units are affected by terrain as follows:

- i. **Woods**. These are impassable to all units.
- ii. **Towns**. These do not restrict the movement of any unit.
- iii. **Marshland and Lakes**. These are impassable to all units.
- iv. **Rivers**. These may only be crossed via bridges or fords.
- v. **Roads**. Units moving by road increase their movement distance by 3" if their entire move is spent on the road. This bonus may not be received if charging.

Moving and Shooting. Units may not shoot if they have moved during the same turn.

Interpenetration. Units may never pass through each other.

Charge Moves. Charges are resolved by moving the attacking unit into contact with its target. They are subject to the following restrictions:

- i. **Turning**. A charging unit may turn once, at the start of its move. This evolution may not exceed 45°.
- ii. **Limited Engagement**. Only one attacking unit may contact each face of the target (these being Front, Left Flank, Right Flank, and Rear).
- iii. **Fighting**. Combat is resolved during the Hand-to-Hand Combat phase.

2. SHOOTING

Only Archers may shoot, the procedure for which is as follows:

Adjudge Field of Fire. Units may only shoot at a single target within 45° of their frontal facing.

Measure Range. Archers have a range of 12".

Assess Casualties. Archers roll a die when shooting, and add 2 to the result. The score gives the number of hits the target acquires, which is modified by the following:

i. **Cover**. Units in woods or towns only suffer half the registered number of hits (any fractions are rounded in favour of the unit shooting).
ii. **Armour**. Men-at-Arms acquire half the number of mandated hits (any fractions should be rounded in favour of the unit shooting).

3. HAND-TO-HAND COMBAT

The procedure for Hand- to-Hand Combat is as follows:

One Sided Combat. Units only inflict casualties during their own player turn.

Assess Casualties. Units roll a die. Men-at-Arms and Levies use the unmodified score; Knights add 2 to the result; and Archers subtract 2. The result gives the number of hits the target acquires, which is modified as follows:

i. **Terrain Advantage**. Defenders in woods, towns, on a hill, or holding a river crossing, only suffer half the indicated number of hits (fractions are rounded in favour of the attacking unit).
ii. **Armour**. Men-at-Arms units acquire half the number of mandated hits (rounding any fractions in favour of the attacking unit).
iii. **Flank or Rear Attacks**. Units engaging the enemy flank or rear inflict double the registered number of casualties.

Movement Within Combat. Hand-to-Hand Combat only concludes with the elimination of one of the contesting sides. Units may however turn to face an attack upon their flank or rear, but only if they are not simultaneously being engaged frontally.

4. ELIMINATING UNITS

Units are eliminated upon the acquisition of 15 hits.

Chapter 8

Pike and Shot Wargaming

The pike and shot period is named after the principal weaponry of infantry units, and covers the age of Renaissance monarchy (1450–1650). The concept of the Renaissance was originally a cultural term referring to the European rediscovery of ancient classical learning during the later fifteenth century; its political expression saw the assertion of monarchical power over that of the nobility.

The eclipse of the feudal aristocracy was made possible by gunpowder, and specifically the development of cannon. For the medieval period had seen the nobility and their cavalry retinues dominate the local countryside from their castles, and owing only a loose allegiance to the monarch, their titular overlord. All this changed with the advent of artillery, for two reasons: firstly, because only the king had enough money to afford a substantial collection of ordnance; and secondly, because the new cannon had the power to destroy any existing castle. As a result, the gunpowder revolution led to the development of Renaissance monarchy, with the growth of state power.

It has to be said that the new monarchies were not especially efficient. They may have been able to dominate their nobility physically, but still needed the aristocracy to staff the new government. A somewhat unwieldy bureaucratic structure was therefore developed, running on the basis of royal patronage. Salaries were not especially high, but the monopolistic nature of offices led to much opportunity for corruption – bribery was rampant.

The military consequence of all this was an inability to finance a permanent centralized army. The nobility could still be relied upon to some extent, but most states recruited mercenary companies on an ad hoc basis as required. These would serve for the duration of a campaign, or until their employer failed to pay their salaries – at which point they deserted the colours, and plundered the countryside. Not that inefficiency made warfare any less prevalent; the development of religious strife following the Protestant Reformation made conflict both common and exceptionally brutal.

Gunpowder weaponry played a particularly significant role on the Renaissance battlefield. Artillery was rather immobile, but made a contribution with a preliminary if somewhat ineffective bombardment of the enemy line (its overall lack of impact explains why units of ordnance do not feature in my wargames rules). Infantry firearms were much more significant; the armour piercing ability of these handguns, later referred to as arquebuses and muskets, made cavalry charges especially perilous, and could cause extreme disorder in all units due both to the physical effect, and the psychological impact of their loud noise.

The new firearms did however have two significant drawbacks. The first of these was the time taken to reload, for even the most efficient infantry handgun could only fire one shot a minute. Units accordingly deployed in deep formations, with the front rank firing, then moving to the rear in order to reload. By the time every rank had fired, the initial front rank would be ready to shoot again. The second drawback lay in the limited amounts of ammunition carried; early cartridges were quite bulky, and most troops only carried twelve. The combined effect of these limitations meant that the musketeers had to be protected by a contingent of pikemen, from which phenomenon the wargaming term of 'Pike and Shot' derived.

Some cavalry units also began to use firearms, following the development of the extremely portable pistol (large handguns could not be used with any effect on horseback). Each man carried up to four of these weapons, and the infantry tactic of having one rank fire at a time, then retiring to reload, was embraced by mounted troops too. It all made for a somewhat sedate method of attack; this could however be more effective than a headlong charge into an unbroken hedge of pikemen

More traditional troop types still existed; I have chosen to depict Pike and Shot warfare by following the precedent of my Ancient wargames rules, and including four distinct varieties of unit.

1. INFANTRY

This class covers those units equipped with a combination of pikemen and musketeers. The proportion of each type varied, although the number of musketeers increased markedly by the end of the period. Wargamers should use their discretion as to how their units are constituted; each type should vary between ⅓ and ⅔ of the unit. Infantry always moved rather slowly, thanks to the unwieldy nature of pikes of up to 24' in length. Their musketry was however quite effective, at least until the ammunition ran out: units could then engage the enemy in hand-to-hand combat, which could often prove to be rather protracted. The pikes were vital in such close quarter contact, and proved especially effective against enemy cavalry – horses were understandably unwilling to throw themselves against a hedge of pikes.

2. SWORDSMEN

This category includes all foot soldiers equipped with swords or axes. They sometimes carried muskets as well, which were always discharged at short range prior to charging the enemy; this can as a result be evaluated as part of hand-to-hand combat in the wargame. The absence of pikes made these units more mobile than Infantry, and short weaponry allowed Swordsmen to inflict fearful execution after the initial impact with pikemen. Conversely, the absence of sustained firearms capability and long mêlée weapons rendered them vulnerable to a cavalry charge. Swordsmen did not habitually feature in all armies of the period, but played a significant role in Iberian, Celtic, and Eastern European warfare.

3. REITERS

This class is named after the German mercenary horsemen who effectively defined it, by virtue of their prominence on many battlefields. These men could be described as pistoleers: they trotted up to enemy formations and discharged their small firearms one rank at a time, continuing to do so until their ammunition ran out. They would then engage in hand-to-hand combat, but always at a rather sedate pace – they relied upon discipline and control rather than a headlong charge. This tended to make their shooting quite effective, but resulted in a lack of impetus and diminished impact in hand-to-hand combat.

4. CAVALRY

The old nobility still believed in the efficacy of shock action, and continued to equip themselves with lances and heavy armour. These horsemen were known as gendarmes, and featured in most sixteenth century armies (their seventeenth century equivalents had less armour and replaced lances with swords, but still relied upon the same tactics). They moved with some rapidity, and were extremely effective in hand-to-hand combat against Swordsmen and Reiters. They were however vulnerable to firearms (which could pierce their armour), and Infantry equipped with pikes, the unyielding nature of which tended to frighten the horses.

The rules for Pike and Shot wargames are quite similar to those for the ancient period, since the broad principles are identical – hand-to-hand combat was decisive in both cases. I have therefore avoided unnecessary repetition of identical concepts in the following discussion of the ideas behind my rules; readers can refer back to Chapter 2 if any further explanation is required.

Movement is resolved in a similar way to the ancient rules, the chief differences concerning the effect of woods and towns upon movement. In the case of the former, only Swordsmen may enter forested areas, given that Infantry units could never negotiate their long pikes through the branches. So far as towns were concerned, these were now villages rather than the hamlets of ancient times; different restrictions must apply as a result. Accordingly, Reiters and Cavalry may neither halt within a town nor occupy it: the greater number of buildings would simply not allow horsemen to fight without dismounting.

The increased importance of firearms in the Pike and Shot era means that shooting has to be treated in a different manner from the ancient period. I therefore allow Infantry and Reiters to fire after they move. This is because they really did just that, as each rank fired and another advanced to take its place. More importantly, allowing firing after movement enables firearms to affect enemy units before the latter may charge. This is because all troops suffered casualties from firearms before charging, no matter how rapidly the victim was able to move: this must be reflected in any wargame.

I have given all firearms a range of 12". This seems extremely odd, since Infantry muskets greatly outranged the Reiters' pistols. The rule does however reflect the situation on the battlefield, where the Reiters would send individual ranks forward with some rapidity, discharge their pistols and then withdraw.

Firing is very effective for as long as Infantry and Reiter units maintain their ammunition supply. The limited nature of the latter must be depicted in the wargame however, and is covered by having each unit throw a die whenever it fires; the ammunition runs out on a score of 1 or 2. This means that a unit has a ⅓ choice of losing its shooting capacity each time it fires – from this point onwards, Infantry and Reiters can only harm the enemy by engaging in hand-to-hand combat. I also do not allow Infantry or Reiters to charge enemy units until their ammunition has been expended. This accurately reflects the historical situation, when units equipped with firearms shot at their potential victims in an attempt to induce disorder, and only charging after they could no longer fire.

Hand-to-hand combat was very similar to the ancient period, in that the engagements tended to be protracted affairs which ended with the elimination of one of the antagonists. The rules are therefore very similar to my ancient wargame; the strengths and weaknesses of different units are reflected by modifying their combat effectiveness according to the nature of the opposing unit, or the terrain in which it is located.

Chapter 9

Pike and Shot Wargames Rules

UNIT TYPES

This game features the unit types of Infantry, Swordsmen, Reiters, and Cavalry, each of which occupies a frontage of between 4 and 6 inches. Any size or scale of figure may be used; wargamers should decide for themselves how many figures constitute a given unit.

SEQUENCE OF PLAY

Each complete turn comprises two player turns. Each wargamer follows the sequence listed below in his or her player turn:

1. Movement
2. Shooting
3. Hand-to-Hand Combat
4. Eliminating Units

1. MOVEMENT

Movement Allowances. Units may move up to the distances listed below during their turn:

Unit Type	Movement Distance
Infantry	6"
Swordsmen	8"
Reiters	10"
Cavalry	12"

Turning. Units turn by pivoting on their central point. They may do so at the start and/or the end of their move.

Terrain. Units are affected by terrain as follows:

i. **Woods.** Only Swordsmen may enter.
ii. **Towns.** Cavalry and Reiters may not end their moves in a town.
iii. **Marshland and Lakes.** These are impassable to all units.
iv. **Rivers.** These may only be crossed via bridges or fords.
v. **Roads.** Units moving by road increase their movement distance by 3" if their entire move is spent on the road. This bonus may not be received if charging.

Moving and Shooting. Infantry and Reiters may shoot at the end of their move.

Interpenetration. Units may never pass through each other.

Charge Moves. Charges are resolved by moving the attacking unit into contact with its target. They are subject to the following restrictions:

i. **Ammunition.** Infantry and Reiters may only charge if they are out of ammunition.
ii. **Turning.** A charging unit may turn once, at the start of its move. This evolution may not exceed 45°.
iii. **Limited Engagement.** Only one attacking unit may contact each face of the target (these being Front, Left Flank, Right Flank, and Rear).
iv. **Fighting.** Combat is resolved during the Hand-to-Hand Combat phase.

2. SHOOTING

Only Infantry and Reiters may shoot, the procedure for which is as follows:

Adjudge Field of Fire. Units may only shoot at a single target within 45° of their frontal facing.

Measure Range. Infantry and Reiters have a range of 12".

Assess Casualties. Units roll a die when shooting. The result gives the number of hits the target acquires, which is modified by the following:

i. **Cover.** Units in woods or towns only suffer half the registered number of hits (any fractions are rounded in favour of the unit shooting).

Check Ammunition. Units roll a second die whenever they fire. If this scores a 1 or 2, the unit has run out of ammunition, and may not fire for the remainder of the game.

3. HAND-TO-HAND COMBAT

The procedure for Hand- to-Hand Combat is as follows:

One Sided Combat. Units only inflict casualties during their own player turn.

Assess Casualties. Units roll a die. Infantry and Reiters use the unmodified score; Swordsmen and Cavalry add 2 to the result. The final score gives the number of hits the target acquires, which is modified as follows:

   i. **Cavalry and Reiters.** These units only inflict half the number of registered hits if attacking Infantry (rounding any fractions in favour of the attacking unit).
   ii. **Swordsmen.** These units only inflict half the number of registered hits if attacking Cavalry or Reiters (rounding any fractions in favour of the attacking unit).
   iii. **Terrain Advantage.** Defenders in woods, towns, on a hill, or defending a river crossing, only suffer half the indicated number of hits (rounding any fractions in favour of the attacker).
   iv. **Flank or Rear Attacks.** Units engaging the enemy flank or rear inflict double the registered number of casualties.

Movement Within Combat. Hand-to-Hand Combat only concludes with the elimination of one of the contesting sides. Units may however turn to face an attack upon their flank or rear, but only if they are not simultaneously being engaged frontally.

4. ELIMINATING UNITS

Units are eliminated upon the acquisition of 15 hits.

Chapter 10

Horse and Musket Wargaming

The Horse and Musket period covers European warfare from 1700–1860. Its characteristic features were the development of regular, disciplined armies; much more potent and flexible infantry; the growth of battlefield artillery; and the revival of cavalry charges.

These military developments stemmed from the disintegration of Renaissance monarchies. These were noted for their inefficiency and confessional strife (see Chapter 8), the nadir of which became apparent during the Thirty Years War (1618–1648). This protracted struggle between Protestant and Catholic states saw most countries run out of money, leading to mercenary armies becoming unpaid – and ravaging every square inch of Germany as a result. This appalling cataclysm understandably led to the desire for something better, and many states gratefully seized upon the opportunities provided by world trade to increase their revenue, which in turn led to more prosperous governments. Economic gains were reinforced by political trends in turn; the age of ecstatic religious enthusiasm had seen such appalling consequences as to promote a very different outlook: the new age of reason relied upon reflection and moderation rather than the passions of fanaticism. It could be argued that the French Revolution and the Napoleonic wars saw a revival of the age of excess; but Europe as a whole did tend to prize the development of the mind over the purity of the soul.

Military innovations coincided with these political developments. The increased state revenues resulted in the recruitment of permanent armies, whose disciplined nature greatly increased their potency. This was especially true of infantry units, which became more mobile and effective thanks to sustained training. They also benefited from two new innovations. The first of these was the flintlock musket, which could allow an average soldier to fire one shot every thirty seconds – fully twice as often as its matchlock predecessor. The cartridges of the new weapon were also much more compact, allowing each man to carry up to sixty; his renaissance counterpart could manage just twelve. This increased firepower provided a significant deterrent effect against those enemy units planning to charge the infantry, which consequence was supported by the invention of the bayonet in the second half of the seventeenth century. The combination of increased firepower and a hedge of cold steel often served to prevent an enemy charge. This rendered pikemen redundant, allowing for more rapid movement as well as increased firepower.

The Enlightenment also saw the development of a new type of infantryman: the Skirmisher. Units of these operated in dispersed formation, relying on rapid movement to keep them out of harm's way. Their open order array allowed Skirmishers to operate in heavily wooded terrain, unlike their close order counterparts. This new breed of light infantry played a subordinate but still effective role on the battlefield: its mobility gave it a great nuisance value.

The development of powerful states saw artillery transformed beyond all recognition. Gunners became much more effectively trained under regular regimentation, which placed them in a good position to take advantage of the new ordnance that was becoming available. The guns were now much lighter and hence more mobile than their Renaissance predecessors, allowing them to perform effectively on the battlefield. Most artillery pieces fired two types of ammunition: solid balls would plough through several enemy ranks at long range; canister ammunition would be resorted to at close proximity – this mode of destruction took the form of packing several musket balls in a thin canister, which burst when the gun was fired, spreading the balls in an effect similar to a modern shotgun. I have not distinguished between the two types of ammunition in my rules, but have instead calculated an average effectiveness for purposes of simplicity.

Cavalry was able to revert to its shock role during the horse and musket period. This was firstly because the pistol was hopelessly outclassed by the new flintlock muskets, rendering a firearms duel somewhat perilous; and secondly because the lack of pikemen gave cavalry a greater chance of success against infantry, despite the deterrent effect of the bayonet. Horsemen made full use of their mobility during the eighteenth and nineteenth centuries – the aim was always to manoeuvre around the enemy, in order to launch a devastating attack upon the flank or rear.

My horse and musket rules have continued the practice of having four different troop classifications, which are listed below:

1. INFANTRY

This category covers all close order foot soldiers equipped with muskets and bayonets. They rely upon their muskets to damage the enemy; it was always theoretically intended that infantry charge enemy units at bayonet point, but most battles degenerated into murderous close range firefights. I accordingly do not allow Infantry units to charge the enemy in these wargames rules.

2. SKIRMISHERS

This class encompasses light infantry operating in dispersed formation. Units of Skirmishers are only half the size of close order foot, and their musketry is correspondingly less effective. Their dispersed formation does however result in rapid movement, and the ability to operate in the sort of difficult terrain that other troops cannot negotiate.

3. ARTILLERY

This covers batteries of ordnance. Artillery has a much longer range than Infantry (48" compared to 12"); its fire is less effective than close order foot, but as potent as that of Skirmishers. Gun batteries are not very mobile: they only move at the speed of Infantry, and may not occupy towns, given that houses obstructed the access of guns and ammunition holders.

4. CAVALRY

This class describes horsemen operating in close order. Cavalry relies entirely upon hand-to-hand combat for its effect, and moves more rapidly than any other category of troops.

The broad principles of these rules are described in Chapter 2; the remainder of this chapter discusses how the specific features of horse and musket warfare are depicted therein.

The movement stipulations depict the mobility of Skirmishers by preventing all other troop types from entering woods; they also allow Skirmishers to move through other friendly units (and vice versa). This means that light infantry are most effective, allowing them to snipe at the enemy from the cover of a wood; or advancing ahead of friendly Infantry, to engage the enemy before the main body of the army arrives.

Towns are now treated as such – they are no longer villages or hamlets. This allows their occupants to enjoy a field of fire of 360°, creating the sort of strongpoints that played a significant part in major battles of the horse and musket period.

The rules for shooting are straightforward, with Infantry enjoying a greater impact than Artillery or Skirmishers as already discussed. The range of 12" for musketry is quite long, but the effect is accurate: it allows for foot units to engage enemy Cavalry before hand-to-hand combat occurs, avoiding the need for allowing the extra complication of having defending troops strike back in a mêlée.

Hand-to-hand combat is simplified by only allowing Cavalry to charge the enemy. This provides a broadly accurate reflection of historical practice (Infantry would only charge an enemy that was on the verge of running away – a situation covered by eliminating the afflicted unit in the rules). I have also simplified hand-to-hand engagements by only allowing a single cavalry unit to charge a given enemy. This reflects the overall nature of horse and musket warfare – charges were brief affairs, without the protracted multiple engagements of preceding centuries. Trying to launch a plurality of units against a single enemy would only have resulted in extreme confusion rather than enhanced effectiveness, which is why such unseemly pile-ups do not feature in these rules.

Cavalry charges can be very potent, once enemy musketry is endured, with flank charges being especially deadly – as was the case historically. Attacks uphill or across a river do however have rather less effect, as is the case when cavalry engage other horsemen; this is because horses tended to shy away from each other if an impact was imminent on the eighteenth- and nineteenth-century battlefield. A cavalry charge is more effective against other units, because the historical defenders tended to flinch from attacking horsemen once the latter had endured musketry and completed their charge. In essence, the effect of any hand-to-hand combat depends entirely upon whichever side showed greater resolution.

Hand-to-hand combat always ends after one assault, with the retreat of the attacking Cavalry. This depicts both the short duration of such engagements during the horse and musket period, and the tendency of Cavalry to be repulsed if failing to destroy their foe.

The horse and musket rules include an optional contingency for the use of square formation. This was often adopted by Infantry units in Napoleonic times in order to repel enemy horsemen; for a hedge of bayonets facing all directions, and backed up by musketry at point blank range, would serve to deter any cavalry charge. For wargaming purposes, squares are immobile (only the most disciplined of units could retain the formation whilst moving), and may not fire (any musketry was reserved for shooting at horsemen who came in close proximity); they are however immune from enemy Cavalry units. Use of the rule for squares allows for combined arms tactics: Cavalry can approach the flank of an enemy Infantry unit, effectively forcing the latter into square formation; the latter can then be assailed by fire from friendly Infantry, Skirmishers and Artillery. One must restate that squares were primarily a feature of Napoleonic warfare, and that wargamers preferring an Enlightenment backdrop for their encounters should not use this rule; also, like any optional rule, it should only play a part in a game if both players agree.

Chapter 11

Horse and Musket Wargames Rules

UNIT TYPES

This game features the unit types of Infantry, Cavalry, Skirmishers, and Artillery. The first three occupy a frontage of 4–6 inches, whereas ordnance is deployed over a width of 2–3 inches. Any size or scale of figure may be used; wargamers should decide for themselves how many figures constitute a given unit.

SEQUENCE OF PLAY

Each complete turn comprises two player turns. Each wargamer follows the sequence listed below in his or her player turn:

1. Movement
2. Shooting
3. Hand-to-Hand Combat
4. Eliminating Units

1. MOVEMENT

Movement Allowances. Units may move up to the distances listed below during their turn:

Unit Type	Movement Distance
Infantry and Artillery	6"
Skirmishers	9"
Cavalry	12"

Turning. Units turn by pivoting on their central point. They may do so at the start and/or the end of their move.

Terrain. Units are affected by terrain as follows:

i. **Woods.** Only Skirmishers may enter.
ii. **Towns.** Only Infantry and Skirmishers may end their move in a town.
iii. **Marshland and Lakes.** These are impassable to all units.
iv. **Rivers.** These may only be crossed via bridges and fords.
v. **Roads.** Units moving by road increase their movement distance by 3" if their entire move is spent on the road. This bonus may not be received if charging.

Moving and Shooting. Units may not shoot if they have moved during the same turn.

Interpenetration. Only Skirmishers may pass through other units (and vice versa).

Charge Moves. Cavalry are the only unit type that may enter Hand-to-Hand Combat. Charges are resolved by moving the attacking unit into contact with their victim. They are subject to the following restrictions:

i. **Turning.** A charging Cavalry unit may turn once, at the start of its move. This evolution may not exceed 45°.
ii. **Limited Engagement.** Only one attacking unit may engage any given defending unit.
iii. **Fighting.** Combat is resolved during the Hand-to-Hand Combat phase.

2. SHOOTING

The procedure for Shooting is as follows (note that cavalry units do not shoot; all other units can).

Adjudge Field of Fire. Units may only shoot at a single unit within 45° of their frontal facing, except for:

i. **Units in Towns.** These have a field of fire of 360°, and may therefore engage any single target in any direction.

Measure Range. Infantry and Skirmishers have a range of 12"; Artillery has a range of 48".

Assess Casualties. Units roll a die when shooting. Infantry use the unmodified score; the result for Skirmishers and Artillery is reduced by 2. The final score gives the number of hits the target acquires, which is modified by the following:

  i. **Cover.** Units in woods or towns only suffer half the registered number of hits (any fractions are rounded in favour of the unit shooting).

3. HAND-TO-HAND COMBAT

The procedure for Hand-to-Hand Combat is as follows:

One Sided Combat. Only the attacking unit inflicts casualties.

Assess Casualties. The attacking cavalry unit rolls a die and adds 2 to the result; this denotes the number of hits acquired by the defender, which are modified by the following:

  i. **Terrain.** If the defending unit occupies a hilltop, it only suffers half the indicated number of hits (rounding any fractions in favour of the attacker).
  ii. **Cavalry Targets.** Defending Cavalry units only acquire half the registered number of hits (rounding fractions in favour of the attacker).
  iii. **Flank or Rear Attacks.** Units engaged in their flank or rear suffer double the normal amount of hits.

Retreat. If failing to destroy the enemy, attacking Cavalry units retreat 6" after the combat is resolved, ending the move facing their erstwhile target.

4. ELIMINATING UNITS

Units are eliminated upon the acquisition of 15 hits.

5. SQUARE FORMATION (OPTIONAL RULE)

Infantry units may move to or from Square formation at the start of the movement phase. The formation has the following effects:

  i. **No Movement.** Units in Square may not move.
  ii. **No Shooting.** Units in Square may not shoot.
  iii. **Protection from Cavalry.** Enemy Cavalry units may not charge an Infantry unit in Square formation.

Chapter 12

Rifle and Sabre Wargaming

This period covers European warfare from 1860–1900, when troops still operated according to the tactical precepts of the Horse and Musket age, but whose capabilities were radically altered by the advent of rifled weapons.

Both warfare and politics saw seismic change thanks to the advent of industrialization. The increasing mechanization of the economy saw the development of an organized working class with its own socialist political agenda – and consequent challenge to traditional aristocratic authority. The old order did however come to realize that although the workers may not have liked the nobility overmuch, they were highly patriotic: this penchant for uniting around the flag led to previously radical nationalist sentiments becoming highly conservative. Thus it was that the Prussian monarchy led the cause of German unification, leading to the unlikely spectacle of the previously restive socialists and liberals acquiescing in the continued political dominance of the old Junker aristocracy. Such trends were exacerbated by the increased militarization of society – and the development of railways allowed for the rapid transportation of large armies, which led to the growth of conscription.

New technology also led to major changes on the battlefield. This was especially true of infantry firearms and artillery ordnance, both of which enjoyed the benefits of effective rifled barrels and loading via the breech, rather than the muzzle. These new developments meant that all guns could now fire much more rapidly, and with much greater accuracy, than the old muzzle loading smoothbore weapons of the Horse and Musket period. Infantry were now in a position to dominate the battlefield, with artillery becoming increasingly more significant too.

Cavalry was especially badly affected by the new rifled weaponry. Horsemen became vulnerable and marginalized; always a conservative arm, the cavalry remained loyal to their sabres, trying to cling to the illusion that traditional shock action still had a place on the battlefield. They consequently refused to adopt the new firearms themselves, and tried instead to launch old fashioned charges – most of which were doomed to failure, with the very occasional exception which only encouraged cavalry to cling to their dangerous illusions. The rifle was now dominant; the sabre still prominent, but obsolescent.

So far as wargames rules are concerned, the Rifle and Sabre period can be treated as essentially the same as the Horse and Musket, with allowances being made for the greater potency of rifled weaponry. The troop types are for example identical, as is the way they operate on the battlefield. The changes lie in respective combat abilities: Infantry, Artillery and Skirmishers all fire with greater effect than their Horse and Musket counterparts; Cavalry conversely functions less effectively. This is the easiest way of reflecting the relative impotence of horsemen: an alternative approach is to make their charges more potent than the rules allow, but to render Cavalry units correspondingly more vulnerable to fire by suffering greater losses than other troops when shot at. The new rules no longer halve the combat result of Cavalry when engaging other horsemen – this has already been accounted for by the reduced combat effect accruing to Cavalry combat in general.

A final change from the Horse and Musket rules is the abolition of the optional rule for Infantry in square formation. This may have been necessary in the age of the smoothbore musket, but was thoroughly redundant with the advent of the breechloading rifle – the extra amount of lead flying around would deter horses almost as much as the square, and allow the Infantry to function much more effectively than they would have done in that tightly packed and immobile formation.

Chapter 13

Rifle and Sabre Wargames Rules

UNIT TYPES
This game features the unit types of Infantry, Cavalry, Skirmishers, and Artillery. The first three occupy a frontage of 4–6 inches, whereas ordnance is deployed over a width of 2–3 inches. Any size or scale of figure may be used; wargamers should decide for themselves how many figures constitute a given unit.

SEQUENCE OF PLAY
Each complete turn comprises two player turns. Each wargamer follows the sequence listed below in his or her player turn:

1. Movement
2. Shooting
3. Hand-to-Hand Combat
4. Eliminating Units

1. MOVEMENT

Movement Allowances. Units may move up to the distances listed below during their turn:

Unit Type	Movement Distance
Infantry and Artillery	6"
Skirmishers	9"
Cavalry	12"

Turning. Units turn by pivoting on their central point. They may do so at the start and/or the end of their move.

Terrain. Units are affected by terrain as follows:

i. **Woods.** Only Skirmishers may enter.
ii. **Towns.** Only Infantry and Skirmishers may end their move in a town.
iii. **Marshland and Lakes.** These are impassable to all units.
iv. **Rivers.** These may only be crossed via bridges and fords.
v. **Roads.** Units moving by road increase their movement distance by 3" if their entire move is spent on the road. This bonus may not be received if charging.

Moving and Shooting. Units may not shoot if they have moved during the same turn.

Interpenetration. Only Skirmishers may pass through other units (and vice versa).

Charge Moves. Cavalry are the only unit type that may enter Hand-to-Hand Combat. Charges are resolved by moving the attacking unit into contact with their victim. They are subject to the following restrictions:

i. **Turning.** A charging Cavalry unit may turn once, at the start of its move. This evolution may not exceed 45°.
ii. **Limited Engagement.** Only one attacking unit may engage any given defending unit.
iii. **Fighting.** Combat is resolved during the Hand-to-Hand Combat phase.

2. SHOOTING

The procedure for Shooting is as follows (note that cavalry units do not shoot; all other units can).

Adjudge Field of Fire. Units may only shoot at a single unit within 45° of their frontal facing, except for:

i. **Units in Towns.** These have a field of fire of 360°, and may therefore engage any single target in any direction.

Measure Range. Infantry and Skirmishers have a range of 12"; Artillery has a range of 48".

Assess Casualties. Units roll a die when shooting. Artillery and Skirmishers use the unmodified score; Infantry add 2 to the result. The final score gives the number of hits the target requires, which is modified by the following:

i. **Cover.** Units in woods or towns only suffer half the registered number of hits (any fractions are rounded in favour of the unit shooting).

3. HAND-TO-HAND COMBAT

The procedure for Hand-to-Hand Combat is as follows:

One Sided Combat. Only the attacking unit inflicts casualties.

Assess Casualties. The attacking cavalry unit rolls a die; this denotes the number of hits acquired by the defender, which are modified by the following:

i. **Terrain.** If the defending unit occupies a hilltop, it only suffers half the indicated number of hits (rounding any fractions in favour of the attacker).
ii. **Flank or Rear Attacks.** Units engaged in their flank or rear suffer double the normal amount of hits.

Retreat. If failing to destroy the enemy, attacking Cavalry units retreat 6" after the combat is resolved, ending the move facing their erstwhile target.

4. ELIMINATING UNITS

Units are eliminated upon the acquisition of 15 hits.

Chapter 14

American Civil War Wargaming

The American Civil War (1861–1865) has always been understandably popular with wargamers in the United States, but has also aroused much interest elsewhere owing to its inherent fascination for the military historian. This is because it can be seen as the first modern war: the significance of mass mobilization of the citizenry, and the primacy of firepower on the battlefield, anticipated the world wars of the twentieth century.

The American Civil War was fought over the issue of the rights of individual states, and stemmed from the very different nature of the northern and southern parts of the United States. Much of the north was industrialized, and relied upon a fluid economy with the free movement of labour; the south was by contrast agrarian, with a static economy and a stratified society – including the institution of slavery. This led to great tension with the north, which was exacerbated as the United States expanded westwards, with the south being terrified that new states would reject slavery, eventually leading to the southern states being outvoted in Congress. The perception that their way of life was undermined led to the southern states attempting to secede from the Union, and war with the north that was determined not to let that happen. The conflict ended with the complete defeat of the Southern Confederacy; and the forcible re-integration of all states within the Union.

The military problems created at the start of the conflict were formidable. Americans had always believed that large armies posed an intrinsic threat to traditional liberties. The regular army was accordingly small; a citizen militia was supposed to rise up and defend the nation in the event of an existential crisis. This view may have had splendid ideological purity and liberal moral rectitude, but resulted in significant problems during the American Civil War, when large numbers of untrained men had to be enlisted very quickly. The existing officer corps found itself in great demand from both sides, but this too created problems: men who had previously commanded a hundred men were told to lead ten thousand. This had some rather unanticipated results, as some previously brilliant officers were found wanting (such as the Union General McClellan); other men with a distinctly mediocre record proved to be inspired commanders at the highest level (the Union General Grant being the most notable example).

The masses of untrained men behaved rather unpredictably on the battlefield, as they inevitably proved incapable of following the complex manoeuvres prescribed in existing drillbooks. The result was that units adopted a rather loose formation, with their undisciplined nature leading to an inability to engage the enemy in hand-to-hand combat. Both infantry and cavalry units would instead indulge in lengthy firefights with the enemy.

The trend towards shooting rather than mêlée was encouraged by new weaponry, as muzzle loading rifles had taken over from the old smoothbore muskets. Rifled barrels allowed much greater accuracy as well as a longer effective range; the invention of a new type of bullet now meant that rifles could be reloaded as quickly as the old weapons with smooth barrels. There has been a great deal of debate over whether new weaponry or undisciplined troops was responsible for the primacy of the firefight; the trend was however incontestable, and my wargame rules reflect it: musketry and artillery is decisive, and hand-to-hand combat banned.

My rules provide for four different troop types, the characteristics of which are outlined below:

1. INFANTRY

The effects of loose order formations are the prime considerations that must be accounted for with foot soldiers in the American Civil War. I therefore allow Infantry units to enter woods, but ban them from engaging in hand-to-hand combat.

2. ZOUAVES

All Infantry units may have been created equally, but some proved to be decidedly more equal than others. 'Zouaves' is a convenient generic term covering all élite infantry: Zouave units were themselves modelled upon European regiments of that designation, and adopted the rather flamboyant dress that was characteristic of the breed. It has to be said that not all Zouave units performed especially well (although many did); they did however see themselves as being an élite, and their presence does allow wargamers to distinguish them by virtue of their distinctive apparel. Less romantic if more historically accurate wargamers may simply dress all foot soldiers in similar rather drab uniforms, and classify individual units as 'Zouaves' (or 'Elite Infantry' if he or she prefers).

Zouave units may not enter hand-to-hand combat, but do move more rapidly than Infantry. Their musketry is also more effective, as befitting their more capable battlefield performance.

3. ARTILLERY

The overall effect of Artillery was roughly the same, irrespective of whether or not it was equipped with smoothbore or rifled ordnance (the former was still more popular and prevalent on the battlefield). Artillery units had a similar effect to their Horse and Musket predecessors, and are treated accordingly in these rules.

4. CAVALRY

Cavalry units showed a distinct inability to charge during the American Civil War, resulting in their effectively being treated as extremely mobile Infantry (although incapable of moving in woods or occupying towns, owing to the inability of horses to operate in such obstructive terrain). Cavalry musketry is however less effective than that of Infantry, reflecting the fact that mounted units were smaller than those on foot, and that some men had to hold the horses of those who dismounted to fire. Cavalry troopers invariably got off their horses to shoot: firing from horseback was never a practical proposition, given that an animal could scarcely be described as a stable firing platform.

The best way of depicting the American Civil War is by effectively using my Horse and Musket wargames rules, but by banning hand-to-hand combat, and allowing Cavalry to shoot. Other special considerations are covered below.

The effects of loose formation have already been mentioned, in that Infantry and Zouave units may move through woods. It could also be argued that loose formation would allow interpenetration, so that units may pass through each other – especially as contemporary drillbooks provided for it, and it was attempted on the battlefield. It did however result in extreme disorder in practice, which is why I do not allow interpenetration in these rules.

The effects of firing have already been discussed. The rules allow Zouaves to shoot more effectively than Infantry, with Artillery and Cavalry conversely being less potent – albeit that ordnance enjoys the benefit of a much longer range. The range for all types of musketry is designed to be that at which firing took effect, rather than the absolute maximum.

Chapter 15

American Civil War Wargames Rules

UNIT TYPES

This game features the unit types of Infantry, Zouaves, Cavalry, and Artillery. The first three occupy a frontage of 4–6 inches, whereas ordnance is deployed over a width of 2–3 inches. Any size or scale of figure may be used; wargamers should decide for themselves how many figures constitute a given unit.

SEQUENCE OF PLAY

Each complete turn comprises two player turns. Each wargamer follows the sequence listed below in his or her player turn:

1. Movement
2. Shooting
3. Eliminating Units

1. MOVEMENT

Movement Allowances. Units may move up to the distances listed below during their turn:

Unit Type	Movement Distance
Infantry and Artillery	6″
Zouaves	9″
Cavalry	12″

Turning. Units turn by pivoting on their central point. They may do so at the start and/or the end of their move.

Terrain. Units are affected by terrain as follows:

i. **Woods.** Only Infantry and Zouaves may enter.
ii. **Towns.** Only Infantry and Zouaves may end their move in a town.
iii. **Marshland and Lakes.** These are impassable to all units.
iv. **Rivers.** These may only be crossed via bridges or fords.
v. **Roads.** Units moving by road increase their movement distance by 3" if their entire move is spent on the road.

Moving and Shooting. Units may not shoot if they have moved during the same turn.

Interpenetration. Units may not pass through each other.

2. SHOOTING

The procedure for shooting is as follows:

Adjudge Field of Fire. Units may only shoot at a single target within 45° of their frontal facing, except for:

i. **Units in Towns.** These have a field of fire of 360°, and may therefore engage any single target in any direction.

Measure Range. Infantry, Zouaves and Cavalry have a range of 12"; Artillery has a range of 48".

Assess Casualties. Units roll a die when shooting. Infantry uses the unmodified score; the roll for Zouaves is increased by 2; and the result for Cavalry and Artillery is reduced by 2. The final score gives the number of hits the target acquires, which is modified by the following:

i. **Cover.** Units in woods or towns only suffer half the registered number of hits (fractions are rounded in favour of the unit shooting).

3. ELIMINATING UNITS

Units are eliminated upon the acquisition of 15 hits.

Chapter 16

Machine Age Wargaming

The 'Machine Age' is a term I coined to cover warfare during 1900–1939, when an industrialized form of military activity based upon weapons technology, became even more significant than the preceding epoch of Rifle and Sabre.

The influence of industrialization led to a profound, indeed symbiotic link between warfare and politics. Otto von Bismarck's success in forging German unity on the back of Prussian military triumph, led to the singularly dangerous belief that warfare was the answer to all social problems. European conservatives thought that socialist tendencies would always be neutralized if the future of the nation was held to be at stake – an analysis that proved to be vindicated at the outbreak of the First World War, when all classes of society proved to be distinctly bellicose.

The Great War should have provided a cautionary lesson for political advocates of homicide on a continental scale: the conflict saw the defeat and disintegration of three great European monarchies – Germany, the Hapsburg Empire, and Russia. Mass ideologies of a decidedly authoritarian stamp did however fill the void, in the shape of fascism and Nazism on the right, and communism on the left. These soon acquired all the militarized qualities and rhetoric of the old regimes, and the conflict engendered by their bellicose outlook led ultimately to the cataclysm of the Second World War.

Any set of wargames rules for the Machine Age must take account of the devastating effects of firepower, which saw the definitive end of hand-to-hand combat. These effects were similar to those of the American Civil War, for all that the latter's recourse to firepower owed more to the indiscipline of its participants, rather than the excellence of its weaponry. Such differences are of great interest to the military historian, but of peripheral significance to the wargame designer: what matters is the fact that the practical consequences were the same; the American Civil War rules can therefore act as a template for any wargaming rendition of the Machine Age. The latter was characterized by rapid-firing magazine rifles, powerful artillery, and the terrifyingly potent machine gun. All this produced a tendency for all troops to effectively adopt skirmish order, in the hope that dispersal would reduce casualties. As with all my other rulesets for this book, I have chosen to depict four common troop types in the Machine Age wargame.

1. INFANTRY

These had exactly the same battlefield effects as their predecessors from the American Civil War. They were equipped with deadly magazine rifles, and adopted dispersed formation in order to reduce the effects of enemy fire.

2. HEAVY INFANTRY

These units represent Infantry with significant support from attached machine guns. This is the best way to reflect the effects of the latter; they were present in most units, but there were never quite enough to go round: it does accordingly make more sense to provide an enhanced combat capability to those few units who engaged the services of a larger than average machine gun allocation. In wargame terms, they enjoy a bonus to all combat dice rolls.

3. ARTILLERY

This category covers light field guns of 75mm calibre, engaging in direct fire at visible targets. They can be treated as having the same range as American Civil War ordnance (they could fire further, but their crew would be unable to see their targets at these longer ranges), but their greater potency means that their combat dice rolling is no longer penalized.

4. CAVALRY

Horsemen still had a role on the battlefields of the Machine Age, albeit solely as mounted infantry equipped with rifles – the days of cavalry charges were well and truly over. Mounted troops generally operated in smaller units than Infantry; this, combined with the fact that some men had to hold the horses when the rest were shooting, means that Cavalry fire is rather less effective than that of Infantry. It can therefore be treated exactly as its American Civil War predecessor, and suffers a reduction in all combat dice rolls accordingly.

Some of you may wonder why I have not included tank units in this wargame. This is mainly because although present, armour was still generally uncommon – albeit spectacularly unwelcome for anyone on the receiving end. Wargamers who absolutely cannot resist including tanks in their games might like to try the following rules: armour has a movement allowance of 9"; it has a firing range of

12" and adds 2 to all combat dice rolls; it also only suffers half casualties when fired upon, thanks to its armoured protection.

One significant change from the American Civil War rules lies in depicting the twentieth century tendency for troops to dig trenches, and for enemy heavy artillery to try to blow them apart. This is simulated by having optional rules for entrenchments and barrages, whereby one side operates a defensive posture by having all troops in open terrain dig entrenchments, which have the same effect as being under cover. The attacker does however enjoy the benefits of a preliminary barrage from heavy ordnance (these weapons have such a long range that they are always deployed off the table; no models need be provided to depict them). The barrage affects 1–3 units, with each victim automatically suffering 1–6 hits. This variability reflects the unpredictability of the artillery barrage: it could potentially decide a battle prior to the assault; it could conversely be so ineffective as to leave the defenders almost entirely intact (as was the case most notoriously with the Battle of the Somme in 1916). The rule for entrenchments and barrages can add a good deal of period flair to Machine Age wargaming but, as is the case with any optional rule, should only be used if both players agree.

Chapter 17

Machine Age Wargames Rules

UNIT TYPES

This game features the unit types of Infantry, Heavy Infantry, Cavalry, and Artillery. The first three occupy a frontage of 4–6 inches, whereas ordnance is deployed over a width of 2–3 inches. Any size or scale of figure may be used; wargamers should decide for themselves how many figures constitute a given unit.

SEQUENCE OF PLAY

Each complete turn comprises two player turns. Each wargamer follows the sequence listed below in his or her player turn:

1. Movement
2. Shooting
3. Eliminating Units

1. MOVEMENT

Movement Allowances. Units may move up to the distances listed below during their turn:

Unit Type	Movement Distance
Infantry, Heavy Infantry and Artillery	6"
Cavalry	12"

Turning. Units turn by pivoting on their central point. They may do so at the start and/or the end of their move.

Machine Age Wargames Rules

Terrain. Units are affected by terrain as follows:

i. **Woods**. Only Infantry and Heavy Infantry may enter.
ii. **Towns**. Only Infantry and Heavy Infantry may end their move in a town.
iii. **Marshland and Lakes**. These are impassable to all units.
iv. **Rivers**. These may only be crossed via bridges or fords.
v. **Roads**. Units moving by road increase their movement distance by 3" if their entire move is spent on the road.

Moving and Shooting. Units may not shoot if they have moved during the same turn.

Interpenetration. Units may not pass through each other.

2. SHOOTING

The procedure for shooting is as follows:

Adjudge Field of Fire. Units may only shoot at a single target within 45° of their frontal facing, except for:

i. **Units in Towns**. These have a field of fire of 360°, and may therefore engage any single target in any direction.

Measure Range. Infantry, Heavy Infantry and Cavalry have a range of 12"; Artillery has a range of 48".

Assess Casualties. Units roll a die when shooting. Infantry and Artillery use the unmodified score; the roll for Heavy Infantry is increased by 2; and the result for Cavalry is reduced by 2. The final score gives the number of hits the target acquires, which is modified by the following:

i. **Cover**. Units in woods or towns only suffer half the registered number of hits (fractions are rounded in favour of the unit shooting).

3. ELIMINATING UNITS

Units are eliminated upon the acquisition of 15 hits.

4. ENTRENCHMENTS AND BARRAGES (OPTIONAL RULE)

The procedure for these is as follows:

Determine Status. Both players roll a die, with the winner deciding whether to attack or defend.

Defender Entrenches. All defending units in the open are entrenched. These units are treated as being under cover.

Attacker Fires Barrage. The attacker targets 1–3 units before the start of the game (roll a die and halve the result, rounding up any fractions). Roll a die for each unit targeted; the victim immediately acquires the indicated number of hits.

Chapter 18

Second World War Wargaming

The Second World War (1939–1945) has always exerted a compelling fascination over many wargamers, given that the unsuccessful attempts by Nazi Germany and Imperial Japan to subjugate their respective continents, gave rise to some of the most memorable campaigns in the history of conflict.

The chief characteristic of twentieth century warfare was the sheer destructive power of the weaponry involved. The infantry were now equipped with rapid firing breechloading rifles, with machine guns providing even more potent support. This rendered the old mass formations obsolete: men on foot could only survive by operating in dispersed array, going to ground whenever they were shot at. Artillery presented an especially potent threat to most troops; its bursting high explosive shells were infinitely more dangerous than the cannonballs of the Horse and Musket period.

The Second World War was also noted for the greater mobility of the troops thanks to the internal combustion engine. Artillery and anti-tank guns were frequently drawn by vehicles rather than horses; some infantry went into battle in armoured personnel carriers; and tanks played a very significant if not necessarily dominant role on the battlefield.

Any set of wargames rules must reflect the wide range of destructive weaponry involved in the Second World War, as well as the greater mobility of all troops. I have selected four main varieties of unit, as with my previous rules; I have however departed from my usual practice by recommending a particular scale of figures. This is because the suggested 1:72 (or 1:76) scale plastic infantry guns and tanks are available from high street toy and model shops. The scale is also compatible with model railway scenery, allowing for the purchase of terrain from the same sources. The four troop types used in my wargame are described below:

1. INFANTRY

The dispersed formation of foot soldiery allows for easy access to woods and towns. Its combat performance is generally adequate, except against tanks. Infantry units (representing around forty real life soldiers), were equipped with some light anti-armour weaponry such as bazookas or small anti-tank guns, but armoured vehicles were generally under limited threat from men on foot, except in the confined spaces of urban areas – here, infantry could get to point blank range and inflict serious damage.

2. MORTARS

This is a generic term covering not only 81mm mortars, but also 75mm light artillery pieces – heavier weapons had such a long range that they could never realistically appear on a small wargames table; they would not be deployed at the low level actions depicted in my scenarios. Mortars would generally be located at some distance from the action, firing at targets that they could not themselves see, but which would be observed by friendly units in radio contact with the mortar unit. Their effectiveness was most pronounced against infantry; high explosive shells could not however pierce the armour of tanks, apart from when a direct hit happened to strike lucky and contact an especially vulnerable part of the vehicle. Mortar units represent about three weapons and their crew.

3. ANTI-TANK GUNS

These very specific weapons are deadly against tanks, but rather ineffective against other units. This is because they shot directly at their targets, rather than at a high trajectory and having shells land on top of any victims. Direct fire was always less potent than a plunging shell when directed against personnel or artillery; anti-tank guns were moreover only supplied with a limited number of high explosive shells – their chief purpose was to engage and destroy enemy tanks. Anti-tank gun models represent about three guns and their crews, along with their towing vehicles (the limited table space does not allow for the depiction of the latter in physical form).

4. TANKS

It is all too common for tanks to enjoy too much prominence and effectiveness on the wargames table; their eye-catching nature can seduce rules designers so much, that an appreciation of their true performance is often lacking. Armoured units perform respectably rather than brilliantly against most units in my rules; they are however deadly when fighting enemy tanks (armoured engagements were historically very short and exceptionally destructive). This reflects the fact that tank guns were essentially anti-tank weaponry in a turret, with similar strengths and weaknesses that apply to units of anti-tank guns; tanks were however also equipped with machine guns, which is why they are more effective against enemy personnel targets than is the case with anti-tank guns. Tank units represent about three vehicles.

The fluidity of movement and dispersed formation is provided for by allowing interpenetration, with all units being able to move through each other. Infantry is rewarded despite its being much slower than some units, by virtue of its capacity to operate in difficult terrain such as woods and towns.

Specific rules for observing targets are a new feature of this set. These are primarily intended to allow other units to locate targets for friendly mortars, which can then be deployed out of harm's way. The observation range is the same as that of most weapons (12"), allowing for easy memorization.

So far as shooting is concerned, dispersed formations allow all units to enjoy a 360° field of fire. Ranges are short: that for mortars is 48", but other units are restricted to 12" – this accounts for the limited visibility of some targets, but also that the effective range of weaponry is not always anywhere near its theoretical maximum. The varying capabilities of troop types against different targets, is covered and effectively explained in the table for casualty assessment; one can instantly see which units are most effective against particular targets, and deploy them accordingly on the wargames battlefield.

Casualties can be reduced by deploying vulnerable units in appropriate terrain, as seen by infantry in woods or towns. It is readily understandable why being under cover would diminish losses, but is less ostensibly apparent why tanks should derive similar benefits when positioned on hilltops. This does however reflect the situation when an armoured unit would be deployed behind the crest of a hill, with only the tank turrets presenting a target. This so-called 'hull down' position allowed armour to engage the enemy with full effect, but only suffer limited casualties in return.

Chapter 19

Second World War Wargames Rules

UNIT TYPES
This game is designed for 1:72 or 1:76 plastic figures and models, which comprise the unit types of Infantry (deployed with 8 figures on a base of 4–6 inches width), Mortars (a weapon with 2 crew on a base 2–3 inches wide), Anti-Tank Guns (a single piece with 2 crew; a base is unnecessary), and Tanks (a single model).

SEQUENCE OF PLAY
Each complete turn comprises two player turns. Each wargamer follows the sequence listed below in his or her player turn:

1. Movement
2. Observation
3. Shooting
4. Eliminating Units

1. MOVEMENT

Movement Allowances. Units may move up to the distances listed below during their turn:

Unit Type	Movement Distance
Infantry and Mortars	6"
Anti-Tank Guns	8"
Tanks	12"

Turning. Units turn by pivoting on their central point. They may do so as often as they like during their move.

Terrain. Units are affected by terrain as follows:

i. **Woods**. Only Infantry may enter.
ii. **Towns**. Only Infantry may end its move within a town.
iii. **Marshland and Lakes**. These are impassable to all units.
iv. **Rivers**. These may only be crossed via bridges and fords.
v. **Roads**. Units moving by road increase their movement distance by 3" if their entire move is spent on the road.

Moving and Shooting. Units may not shoot if they have moved during the same turn.

Interpenetration. Friendly units may pass through each other freely.

2. OBSERVATION

Units may only shoot at units they can see (with one exception: see rule (2c) below). The following rules outline who can see whom:

Observation Ranges. Units may only be observed up to a range of 12".

Line of Sight. Target observation is blocked by hills, woods, towns, and other enemy units.

Indirect Fire. Mortars may shoot at targets they cannot see. The victim must however be observed by another friendly unit.

3. SHOOTING

Units have a field of fire of 360°. The procedure for shooting is as follows:

Measure Range. Infantry, Anti-Tank Guns and Tanks have a range of 12"; Mortars have a range of 48."

Assess Casualties. Units roll a die when shooting, the score of which is modified according to the table below:

	Defending Unit			
Unit Shooting	*Infantry*	*Mortars*	*Anti-Tank Guns*	*Tanks*
Infantry	0	0	0	−2
Mortars	+2	0	0	−2
Anti-Tank Guns	−2	−2	−2	+2
Tanks	0	0	0	+2

The result indicates the number of hits the target suffers, unless it benefits from any of the following:

i. **Cover.** Infantry in woods or towns only acquire half the registered number of hits (fractions are always rounded in favour of the unit shooting).

ii. **Hilltops.** A Tank unit deployed on a hilltop only suffers half the indicated level of hits (fractions are rounded up in favour of the unit shooting).

4. ELIMINATING UNITS

Units are eliminated upon the acquisition of 15 hits.

Chapter 20

Wargame Scenarios

There is a paradox at the heart of wargaming, in that many players are absolutely and rightly fascinated by finding the right set of rules, but pay far less attention to the type of battle (or scenario) which they play. All too many wargamers will acquire many different rulebooks, examine all facets of their contents, and have very definite opinions upon their veracity – and confine their scenario to the traditional pitched battle. This involves a rather contrived encounter featuring armies intended to be absolutely balanced fighting over a wargames battlefield whose terrain favours neither side. This sort of encounter may be ideal for a wargaming competition, but soon becomes rather sterile. More importantly, the pitched battle scenario is intrinsically implausible, given that the whole point of historical generalship was to force the enemy to fight at a disadvantage, rather than to risk one's all on an equal engagement.

The key to any rewarding wargame is therefore an imaginative scenario. I have accordingly included thirty different games in this chapter, which can be fought using any of the rulesets included in this book. All are designed to be fought on small tables of 3' x 3', allowing for accessible encounters in all households; each can be played in one hour. Maps are provided with each scenario in order to facilitate their re-creation on the tabletop: each square on the map represents an area of 12" x 12".

Each scenario has the same format. Most of the commentary provides essential information (such as troop deployment, reinforcement schedules, special rules and victory conditions). Each encounter also includes a description of the situational context, allowing a wider backdrop to the game: this allows players to see that the scenario has a wider strategic purpose, adding character to each encounter.

The sizes of each army are the same in nineteen of the thirty scenarios, but variety is always provided by varying the composition of each. Generals invariably had to operate with the troops they were allocated, rather than those with which they would necessarily prefer to act. This doubtless regrettable if historically accurate fact is accounted for by the following mechanism: players must roll a die and consult the relevant table below to ascertain the composition of his or her army (if identical armies are generated, players should re-roll their dice until distinct forces are created):

Table 1: Armies With 6 Units

	Unit Type			
Die Roll	Infantry Knights	Archers Warband Reiters Artillery Mortars	Skirmishers Levies Swordsmen Zouaves Heavy Infantry Anti-Tank Guns	Cavalry Men-at-Arms Tanks
1	3	2	0	1
2	3	1	2	0
3	3	0	1	2
4	4	1	0	1
5	4	1	1	0
6	4	0	1	1

Table 2: Armies With 4 Units

	Unit Type			
Die Roll	Infantry Knights	Archers Warband Reiters Artillery Mortars	Skirmishers Levies Swordsmen Zouaves Heavy Infantry Anti-Tank Guns	Cavalry Men-at-Arms Tanks
1	2	1	0	1
2	2	1	1	0
3	2	0	1	1
4	3	1	0	0
5	3	0	1	0
6	3	0	0	1

Table 3: Armies With 3 Units

Die Roll	Unit Type			
	Infantry Knights	Archers Warband Reiters Artillery Mortars	Skirmishers Levies Swordsmen Zouaves Heavy Infantry Anti-Tank Guns	Cavalry Men-at-Arms Tanks
1	1	1	0	1
2	1	1	1	0
3	1	0	1	1
4	2	1	0	0
5	2	0	1	0
6	2	0	0	1

The final part of each scenario describes its inspiration. This is a vital part of any game, for all writers should always acknowledge their sources. My scenarios are usually derived either from great historical battles, or games devised by eminent wargames writers. I have invariably changed and abbreviated the original scenarios to allow the re-creation of the essential situation on the wargame table. I have included a list of further reading, for any player who wishes to find out more about my sources.

SCENARIO 1: PITCHED BATTLE (1)

SITUATION
Two armies are facing up to each other over a symmetrical battlefield.

ARMY SIZES
Both armies have 6 units.

DEPLOYMENT
1. The Red army deploys first, within 6" of the northern table edge.
2. The Blue army deploys second, within 6" of the southern table edge.

REINFORCEMENTS
There are no reinforcements in this scenario.

SPECIAL RULES
No special rules apply to this scenario.

GAME LENGTH AND TURN ORDER
This scenario lasts 15 game turns. The Red player goes first in each turn.

VICTORY CONDITIONS
The army which eliminates the greatest number of enemy units is the victor.

INSPIRATION
The situation for this scenario was provided by the Battle of Ceresole (1544), when the French army defeated the forces of the Holy Roman Empire. It has to be said that totally symmetrical battlefields like this are most uncommon in periods after Ancient times (when battles frequently occurred on flat plains). They are however rather more prevalent in wargames tournaments: most competitive encounters are predicated on the notion of contrived equality both in terms of army composition and topographical layout. Such encounters can be very enjoyable, but could be said to lack variety.

FURTHER READING
The following books provide fine accounts of the Battle of Ceresole:

Featherstone, Donald, *Wargaming: Pike-and-Shot* (David and Charles, 1977) (pp. 39–46).
Oman, Sir Charles, *A History of the Art of War in the Sixteenth Century* (Greenhill Books, 1987; originally 1937) (pp 229–243).

Wargame Scenarios 67

SCENARIO 2: PITCHED BATTLE (2)

SITUATION
Two armies are arrayed for battle. Both are aiming to secure the strategically vital objectives of the hill and the crossroads.

ARMY SIZES
Both armies have 6 units.

DEPLOYMENT
1. The Red army deploys first, within 6" of the northern table edge.
2. The Blue army deploys second, within 6" of the southern table edge.

REINFORCEMENTS
There are no reinforcements in this scenario.

SPECIAL RULES
No special rules apply to this scenario.

GAME LENGTH AND TURN ORDER
This scenario lasts 15 game turns. The Red player goes first in each turn.

VICTORY CONDITIONS
Victory is secured by controlling both the hill and the crossroads at the end of the game.

INSPIRATION
This scenario represents a variation of the preceding game. In this version of the classic pitched battle, the goal is not simply to destroy the enemy, but is instead to secure two vital pieces of terrain. This concept does not derive from any specific historical encounter, but has instead been inspired by examples from classic wargames books.

FURTHER READING
The following wargames books contain seminal encounters of this scenario type:

Grant, Charles, *The Ancient War Game* (A and C Black, 1974) (pp. 51–59).
Young, Brig. P and Lt. Col. J. P. Lawford, *Charge!* (Athena Books, 1986; originally 1967) (pp. 29–39).

Wargame Scenarios 69

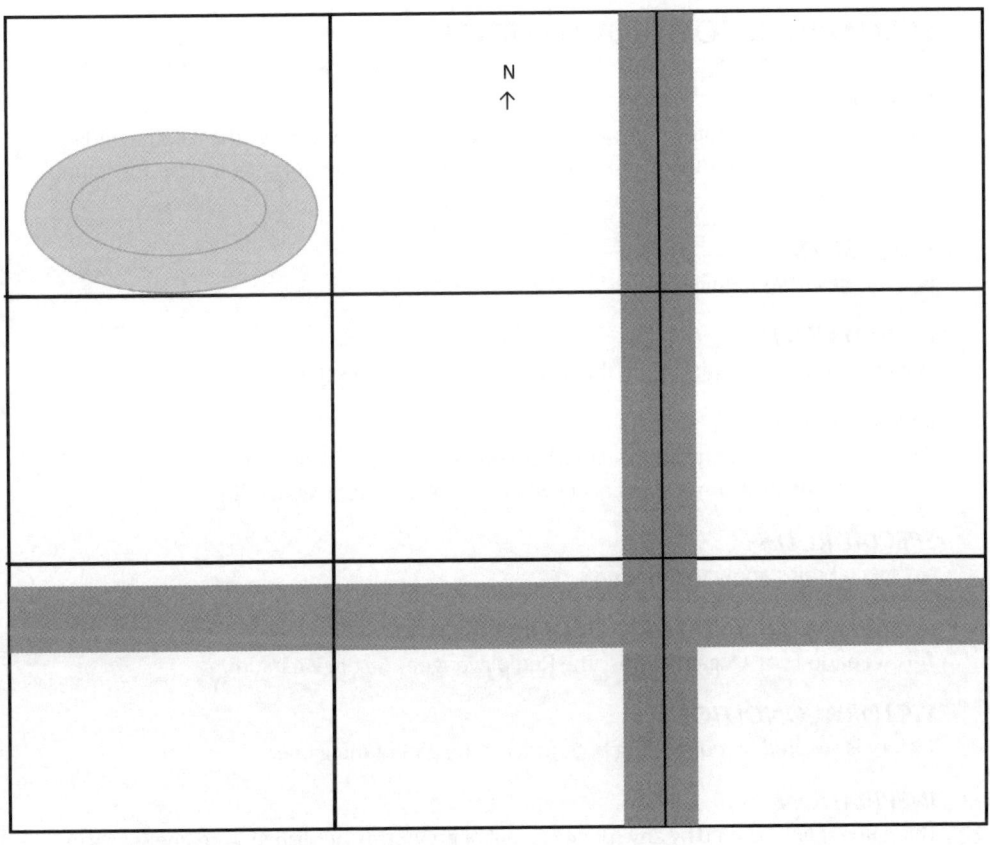

SCENARIO 3: CONTROL THE RIVER

SITUATION
The Red and Blue armies represent portions of much larger forces. Their commanding generals have ordered them to seize two strategic river crossings, as a base for future operations.

ARMY SIZES
Both armies have 6 units.

DEPLOYMENT
The armies are not deployed on the table at the start of the scenario.

REINFORCEMENTS
Turn 1: (a) The Red army arrives anywhere on the northern table edge.
(b) The Blue army appears anywhere on the southern table edge.

SPECIAL RULES
No special rules apply to this scenario.

GAME LENGTH AND TURN ORDER
This scenario lasts 15 game turns. The Red player goes first in each turn.

VICTORY CONDITIONS
Victory is secured by controlling both fords at the end of the game.

INSPIRATION
This game is not inspired by any specific event or previously published wargame scenario, but owes a great deal to traditional wargames, where the control of rivers provides the focus for an interesting game. I also wanted the players to consider themselves as being part of a larger army – this particular engagement may be only a minor affair, but success in securing the river crossings would give the overall army commander a crucial advantage in any campaign.

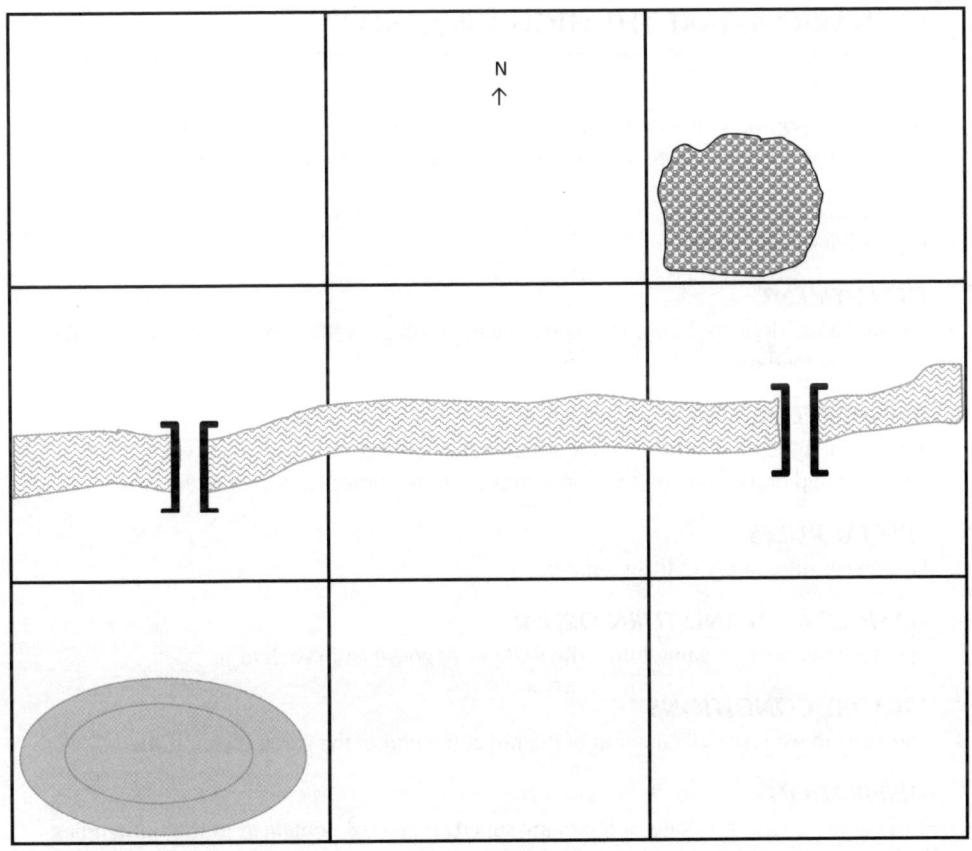

SCENARIO 4: TAKE THE HIGH GROUND

SITUATION
An isolated portion of the Red army occupies a strategic hill. The Blue general has noted this, and has set out to seize the hilltop before enemy reinforcements arrive.

ARMY SIZES
Both armies have 6 units.

DEPLOYMENT
The Red army deploys 2 units on the hill, facing south. No Blue units are deployed at the start of this scenario.

REINFORCEMENTS
Turn 1: The Blue army arrives on the southern table edge.
Turn 2: 4 Red units arrive on the northern table edge, either on or east of the road.

SPECIAL RULES
No special rules apply to this scenario.

GAME LENGTH AND TURN ORDER
This scenario lasts 15 game turns. The Red player goes first in each turn.

VICTORY CONDITIONS
The army in exclusive occupation of the hill at the end of the game is victorious.

INSPIRATION
This game derives from one of the many superb scenarios contained in the outstanding book referred to below. I have modified the terrain to account for the particular conditions and limitations of a small wargames table, but the core concept remains intact – an army attempting to seize a vital objective before enemy reinforcements arrive.

FURTHER READING
The original inspiration for this game can be found in the following book:

Grant, Charles Stewart, *Scenarios for Wargames* (Wargames Research Group, 1981) (pp. 40–41).

Wargame Scenarios 73

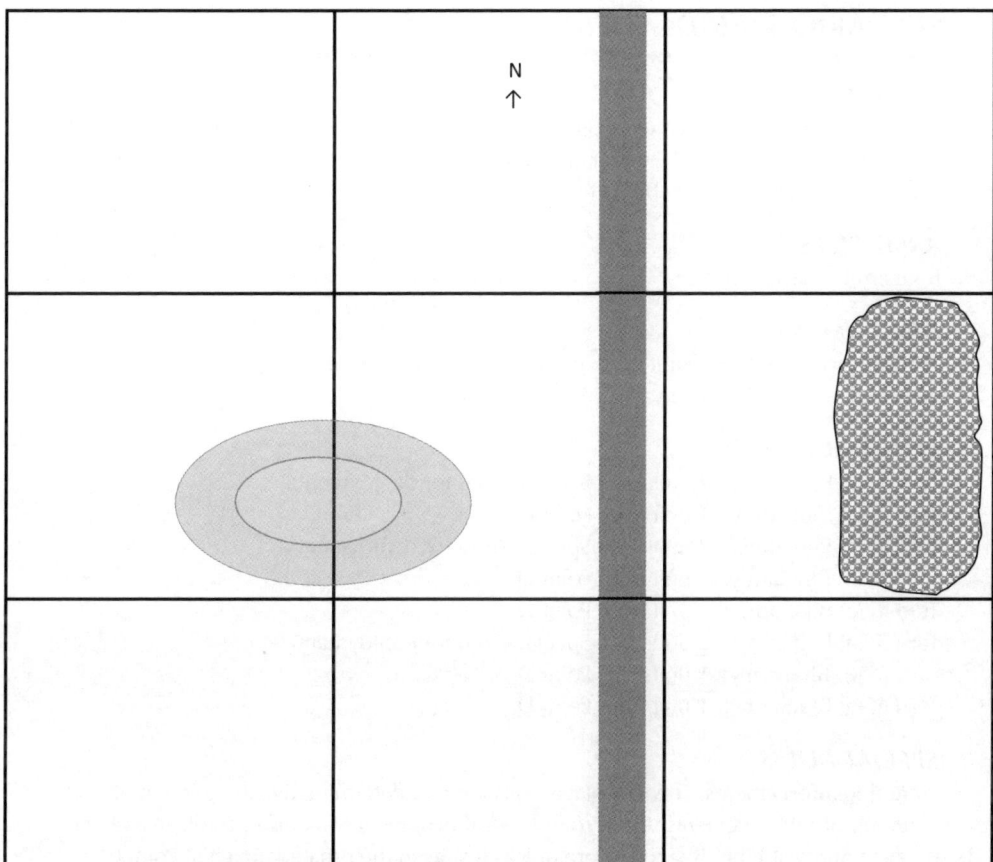

SCENARIO 5: BRIDGEHEAD

SITUATION
The Blue army has discovered a river crossing in Red territory, and is aiming to secure it. The Red general is frantically attempting to mobilize every available unit, in order to stop the enemy bridgehead from being formed.

ARMY SIZES
Both armies have 6 units.

DEPLOYMENT
The Blue army deploys 1 unit north of the river, within 6" of the bridge. No Red units are deployed at the start of this scenario.

REINFORCEMENTS
Turn 1: (a) Red army: 2 units (see Special Rules for deployment).
Turn 2: (a) Blue army: 1 unit from Point D.
Turn 3: (a) Red army: 2 units (see Special Rules for deployment).
      (b) Blue army: 1 unit from Point D.
Turn 4: (a) Blue army: 1 unit from Point D.
Turn 5: (a) Red army: 2 units (see Special Rules for deployment).
      (b) Blue army: 1 unit from Point D.
Turn 6: (a) Blue army: 1 unit from Point D.

SPECIAL RULES
1. **Red Reinforcements**. The Red player rolls a die to determine the arrival point of each group of reinforcements. On a roll of 1–2 they appear at Point A; a roll of 3–4 sees them arrive at Point B; and a score of 5–6 results in their materializing at Point C.

GAME LENGTH AND TURN ORDER
This scenario lasts 15 game turns. The Red player goes first in each turn.

VICTORY CONDITIONS
Victory is achieved by there being no enemy units on the north bank of the river, within 12" of the bridge.

INSPIRATION
This game was derived from another of Charles Stewart Grant's splendid scenarios. I especially wanted to explore the idea of both armies arriving piecemeal, with the added complication of a disorganized Red defence, as the Red general responds to the unwelcome surprise of the Blue incursion.

FURTHER READING
The inspiration for this game can be found in the following book:

Grant, Charles Stewart, *Scenarios for Wargames* (Wargames Research Group, 1981) (pp. 46–47).

Wargame Scenarios 75

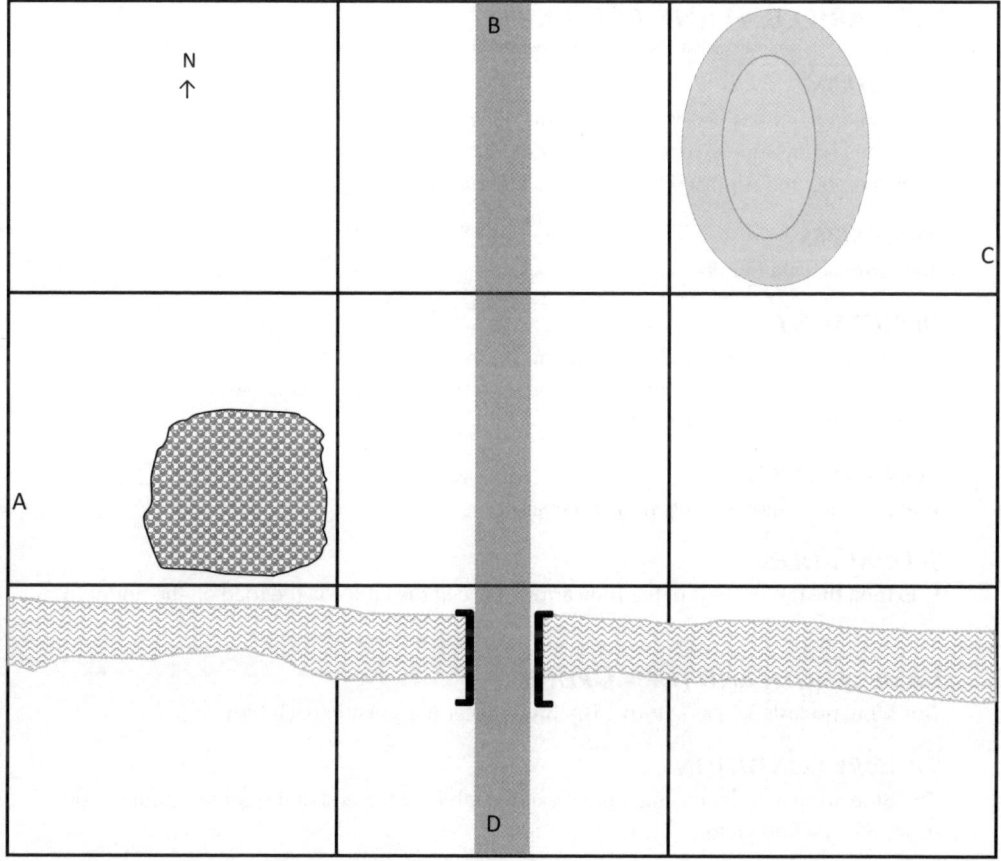

SCENARIO 6: FLANK ATTACK (1)

SITUATION
The Blue general has ordered his army to burst through a small Red blocking force and seize the enemy supply base. Unfortunately for the Blue army, the Red general has seen what is afoot, and is poised to launch his army against the Blue flank.

ARMY SIZES
Both armies have 6 units.

DEPLOYMENT
1. Red army: (a) 2 units in Zone 1, facing south.
   (b) 4 units within 12" of the eastern table edge, facing west.
2. Blue army: in Zone 2, facing north.

REINFORCEMENTS
There are no reinforcements in this scenario.

SPECIAL RULES
1. **Exiting the table**. Units of the Blue army may exit the table via the road on the northern table edge.

GAME LENGTH AND TURN ORDER
This scenario lasts 15 game turns. The Blue player goes first in each turn.

VICTORY CONDITIONS
The Blue army wins by having 3 units exit the table by the end of the game. Failure to do so results in a Red victory.

INSPIRATION
This game is derived from the Battle of Salamanca (1812), which saw the Duke of Wellington's Anglo-Portuguese army turn the tables on a French move against the Duke's line of communications. This scenario sees Wellington's army (Red) seeking to re-enact the achievement. Players should note that the Blue army still has the initiative, which is why they have the first turn; the Red outflanking move must be launched quickly and decisively in order to prevent excessive pressure against the contingent in Zone 1.

FURTHER READING
Accounts of the Battle of Salamanca can be found in the following works:

Lipscombe, Col. Nick, *The Peninsular War Atlas* (Osprey, 2010) (pp. 256–273).
Weller, Jac, *Wellington in the Peninsula, 1808–1814* (Greenhill Books 1992; originally 1962) (pp 206–230).

Wargame Scenarios 77

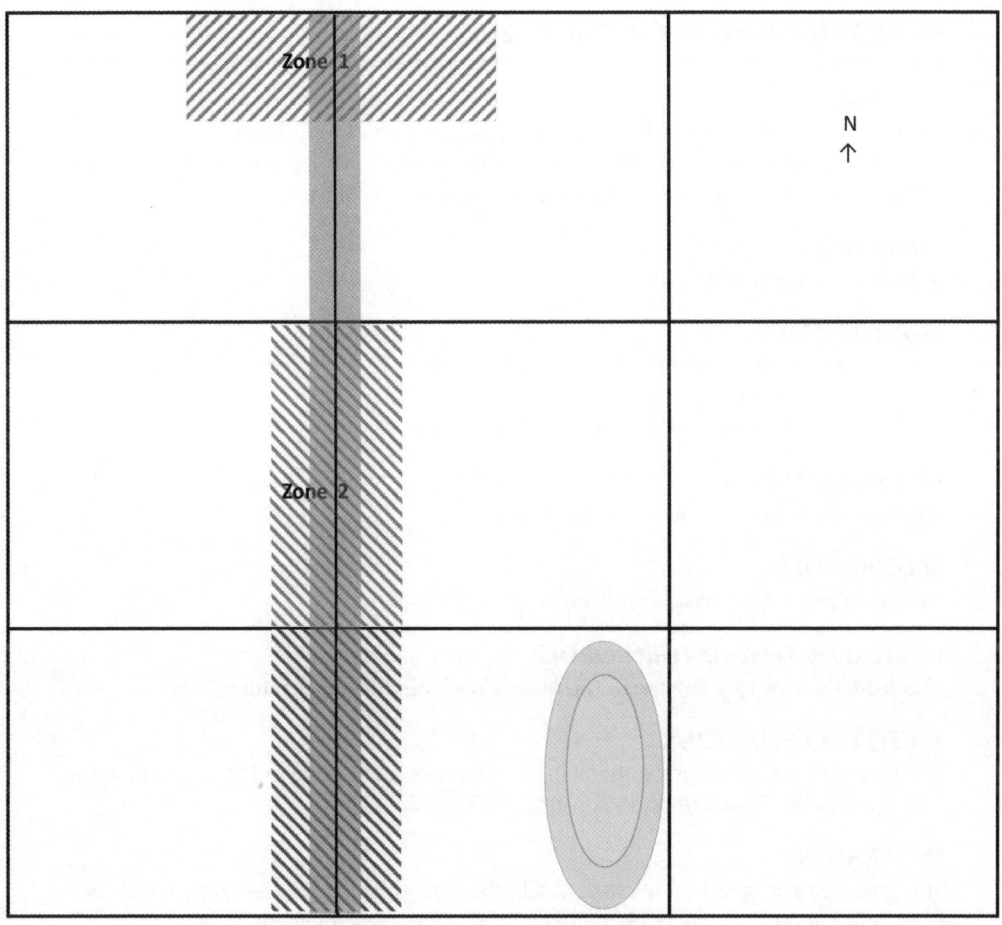

SCENARIO 7: FLANK ATTACK (2)

SITUATION
The Red army is deployed on a dominant hill, expecting an attack from the south. Unfortunately for the Red general, his Blue counterpart has taken advantage of some concealed ground, and sent the bulk of his army on a march around the Red force's left flank.

ARMY SIZES
Both armies have 6 units.

DEPLOYMENT
1. Red army: deploys on the large hill, facing south.
2. Blue army: (a) 2 units on the small hill, facing north.
              (b) 4 units in Zone 1, facing west.

REINFORCEMENTS
There are no reinforcements in this scenario.

SPECIAL RULES
No special rules apply to this scenario.

GAME LENGTH AND TURN ORDER
This scenario lasts 15 game turns. The Blue player goes first in each turn.

VICTORY CONDITIONS
The Blue player secures victory by being in exclusive occupation of the large hill at the end of the game. Failure to do so constitutes a Red win.

INSPIRATION
This game was inspired by another of Charles Grant's splendid scenarios, which was itself based upon King Frederick the Great of Prussia's masterly manoeuvre at the Battle of Leuthen (1757), which took his Austrian enemies by complete surprise and led to one of the greatest victories in military history.

This scenario differs slightly in character from the preceding game. That saw the outflanked player with significant attacking obligations; this scenario sees the defender with no other task than mere survival.

FURTHER READING
The inspiration for this scenario can be found in the following book:

Grant, Charles Stewart, *Scenarios for Wargames* (Wargames Research Group, 1981) (pp. 34–35).

Readers who wish to discover the events of the Battle of Leuthen will find fine accounts in these books:

Fuller, Maj. Gen. J. F. C. *The Decisive Battles of the Western World (Volume 2)* (SPA Books,1994; originally 1954) (pp. 207–215).
Grant, Charles S, with Charlie and Natasha Grant, *Wargaming in History Volume 4* (Ken Trotman, 2011) (pp. 87–101).

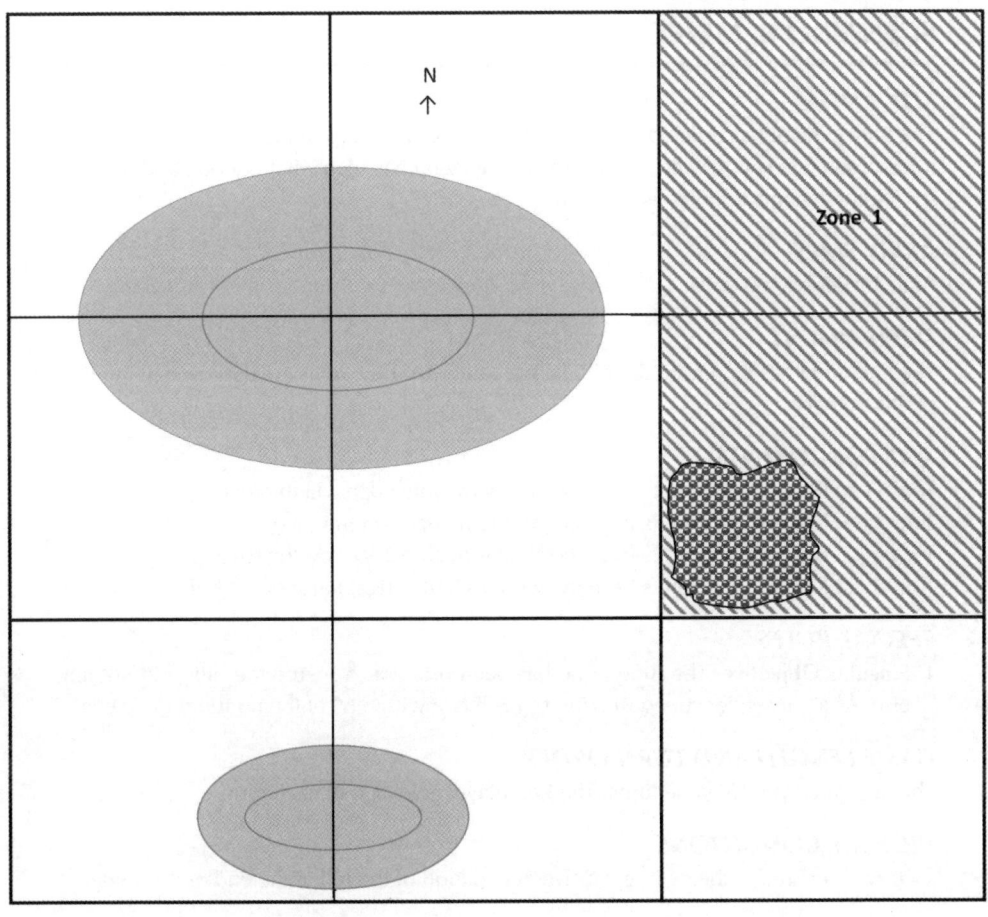

SCENARIO 8: MÊLÉE

SITUATION
The Blue general has ordered a surprise attack upon a strategic hilltop, with an additional force ordered to support the assault. The Red general has seen what is afoot, and has sent for reinforcements to hold his position.

ARMY SIZES
Both armies have 6 units.

DEPLOYMENT
The Red army has 2 units on the hill, facing south. No Blue units are deployed at the start of this scenario.

REINFORCEMENTS
Turn 1: (a) Blue army: 3 units from the southern table edge, via the road.
Turn 3: (a) Red army: 2 units from the northern edge, via the road.
Turn 4: (a) Blue army: 3 units from the southern table edge, via the road.
Turn 6: (a) Red army: 2 units from the western table edge, north of the hill.

SPECIAL RULES
1. **Singular Objective**. The Blue army has been ordered to secure the hill, and nothing else. As a consequence, no Blue unit may move within 6" of the northern table edge.

GAME LENGTH AND TURN ORDER
This scenario lasts 15 game turns. The Red player goes first in each turn.

VICTORY CONDITIONS
Victory is secured by being in exclusive occupation of the hill at the end of the game.

INSPIRATION
This game was derived from the Battle of Lundy's Lane (1814), or more precisely, from a superb book of wargame scenarios by Stuart Asquith devoted to depicting the War of 1812 (between Britain and America) on the tabletop. The idea of a confused engagement around a prominent objective invariably gives rise to exciting wargames, which is why it is included here.

FURTHER READING
An account of, and wargame scenario for, the Battle of Lundy's Lane can be found in the following book:

Asquith, Stuart, *Scenarios for the War of 1812* (Partizan Press, 2010) (pp. 67–71).

Wargame Scenarios 81

SCENARIO 9: DOUBLE DELAYING ACTION

SITUATION
The Blue and Red supreme commanders are occupied in a decisive battle five miles north of our engagement. Both our generals have been ordered to reinforce their superiors' armies, with the Blue general also expected to seize the town as a base for future operations.

ARMY SIZES
Both armies have 6 units.

DEPLOYMENT
The entire Red army is positioned anywhere north of the river. No Blue units are deployed at the start of this scenario.

REINFORCEMENTS
Turn 1: (a) Blue army: all 6 units appear from the southern table edge.

SPECIAL RULES
1. **Exiting the Table**. Both armies may leave the table via the road on the northern edge.
2. **Mandatory Exits**. The Red army must withdraw 1 unit by the end of Turns 4, 8, and 12 (making a total of 3 units exiting the table).
3. **Defensive Posture**. The Red army may never move south of the river.

GAME LENGTH AND TURN ORDER
This scenario lasts 15 game turns. The Blue player goes first in each turn.

VICTORY CONDITIONS
The Blue player wins if either of the following apply:

(a) The Red army fails to withdraw units as mandated by special rule number 2.
(b) The Blue army controls the town and exits 2 units from the table by the end of the game.

INSPIRATION
This fascinating problem of a double delaying action occurred at the Battle of Wavre, fought on the same day as the Battle of Waterloo (1815). The French commander at Wavre, Marshal Grouchy, was ordered to prevent the Prussians under Marshal Blücher from reinforcing the Duke of Wellington. Grouchy's failure does not make the situation any less interesting as a wargame scenario.

FURTHER READING
Wavre is often neglected in accounts of the Waterloo campaign, but one shining exception to this unfortunate omission can be found in the following book:

Hofschröer, Peter, *1815 the Waterloo Campaign: The German Victory* (Greenhill Books, 1999) (pp.154–172).

An interesting attempt at adopting the Battle of Wavre to a wargames scenario set in the eighteenth century can be found in:

Grant, Charles S, *The Wolfenbüttel War* (Partizan Press, 2012) (pp. 42–52).

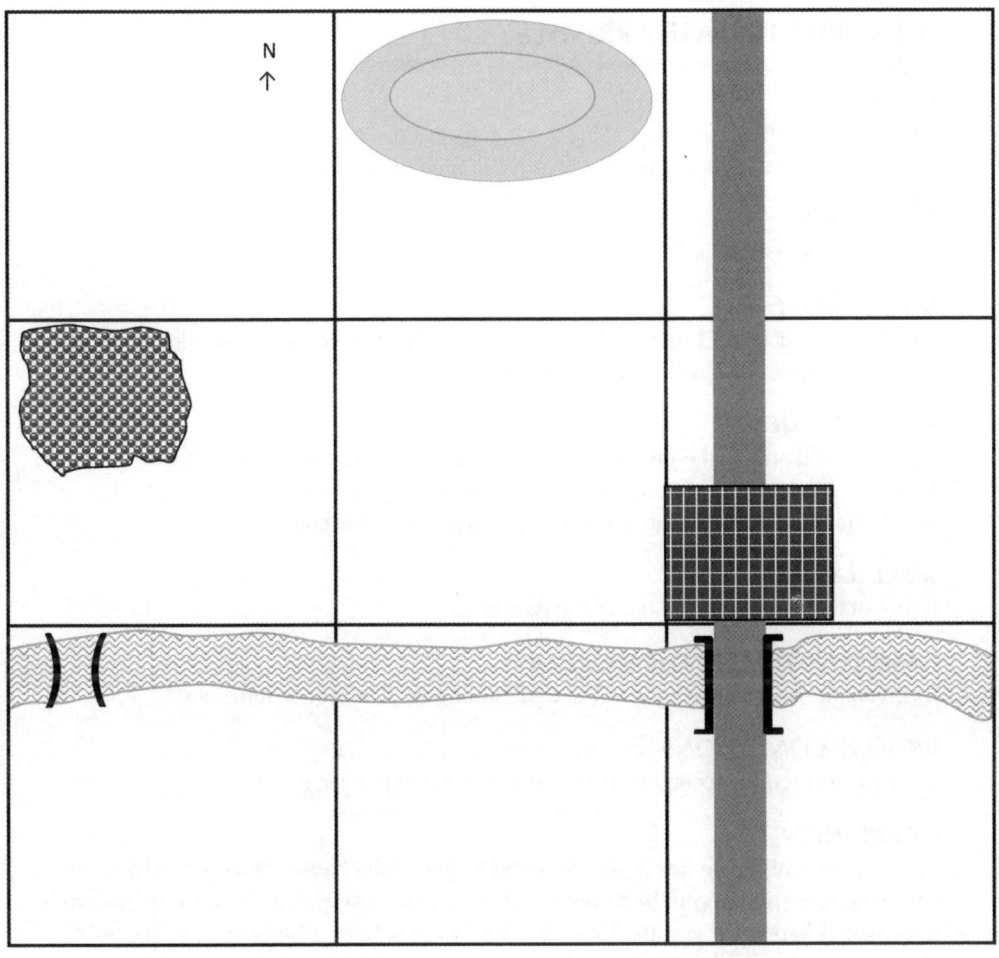

SCENARIO 10: LATE ARRIVALS

SITUATION
A portion of the Blue army is making a stand against the advancing Red force. The Blue general hopes that reinforcements will arrive in time to repel the enemy.

ARMY SIZES
Both armies have 6 units.

DEPLOYMENT
The Blue army deploys 2 units within 24" of the southern table edge and/or in the wood. No Red units are deployed at the start of the game.

REINFORCEMENTS
Turn 1: (a) Red army: all 6 units enter via the road on the northern table edge.
Turn 5: (a) Blue army: 2 units arrive from the southern table edge.
Turn 10: (a) Blue army: 2 units arrive from the southern table edge.

SPECIAL RULES
1. **Mountainous terrain**. The hill is impassable.

GAME LENGTH AND TURN ORDER
This scenario lasts 15 game turns. The Red player goes first in each turn.

VICTORY CONDITIONS
Victory is secured by occupying the town at the end of the game.

INSPIRATION
The terrain for this game was loosely inspired by that of the Battle of Gitschin (1866); the rest of the scenario concept bears little relation to that engagement. I wanted to explore what would happen if an attacking force had to penetrate a bottleneck to secure its objective – and if the defender's forces only arrived in a piecemeal manner.

FURTHER READING
Accounts of the Battle of Gitschin can be found in:

Barry, Quintin, *The Road to Königgrätz* (Helion, 2010) (pp. 267–289).
Craig, Gordon A, *The Battle of Königgrätz* (Weidenfeld and Nicolson, 1965) (pp. 87–91).

A wargame scenario for Gitschin is covered by:

Weigle, Bruce, *1866* (Medieval Miscellanea, 2010) (pp. 57–59).

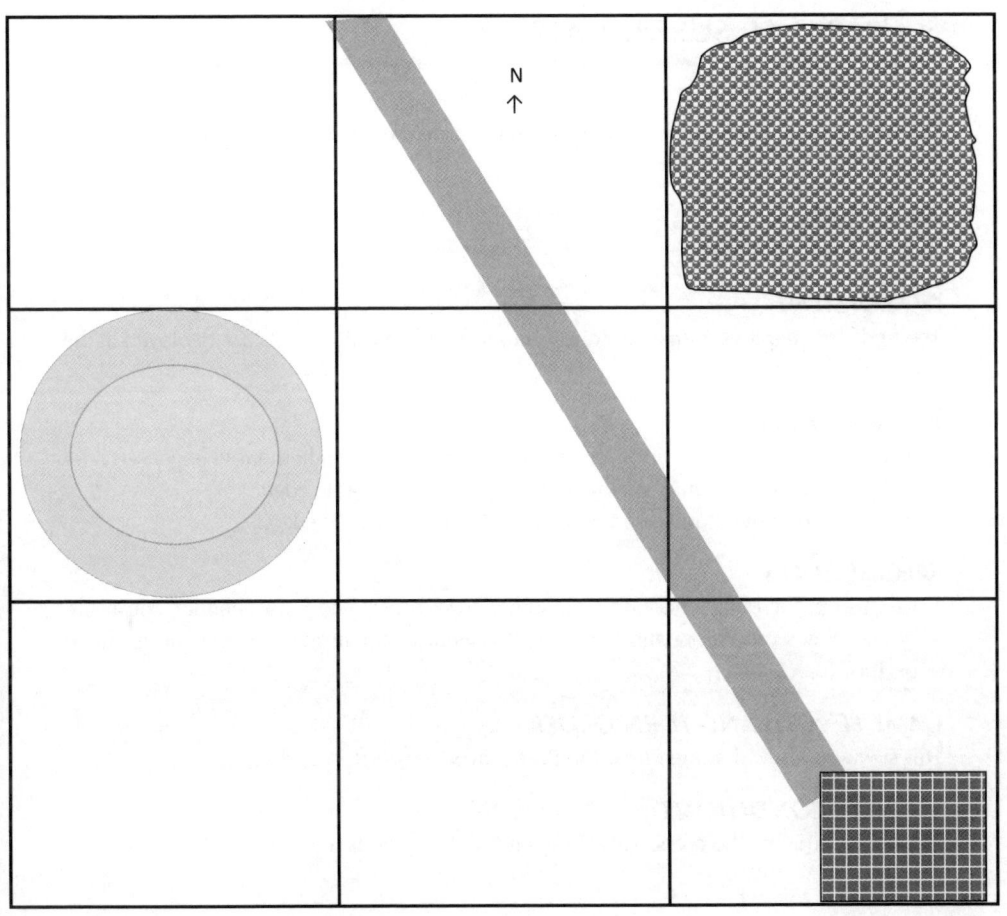

SCENARIO 11: SURPRISE ATTACK

SITUATION
The Blue general has launched a surprise attack, with the intention of capturing a strategic crossroads.

ARMY SIZES
Both armies have 6 units.

DEPLOYMENT
The Red army deploys 2 units in Zone 1, facing south. No Blue units are deployed at the start of this game.

REINFORCEMENTS
Turn 1: (a) Blue army: all 6 units arrive via the road on the southern table edge.
Turn 3: (a) Red army: 2 units via the road on the northern table edge.
Turn 9: (a) Red army: 2 units via the road on the western table edge.

SPECIAL RULES
1. **Confusion**. The initial encounter is assumed to be unexpected, with neither side aware of the other's precise location or strength. Accordingly, no charges may be declared on Turn 1.

GAME LENGTH AND TURN ORDER
This scenario lasts 15 game turns. The Blue player goes first in each turn.

VICTORY CONDITIONS
The side occupying the crossroads at the end of the game is the victor.

INSPIRATION
This game is derived from the Battle of Quatre Bras (1815), fought during the Waterloo campaign. The real engagement was a closely contested and bitter struggle between Marshal Ney's French (the Blue army) and the Duke of Wellington's Anglo-Allied forces (the Red army) – the tabletop encounter should prove most exciting, as the Red army desperately tries to maintain its position.

FURTHER READING
The books listed below provide fine accounts of Quatre Bras. The first title goes into positively microscopic detail from start to finish, which is why no specific page references are given; the second book offers a concise account:

Robinson, Mike, *The Battle of Quatre Bras 1815* (The History Press, 2009).
Uffindell, Andrew, *The Eagle's last Triumph* (Greenhill Books, 2006) (pp. 121–152).

An adaptation of Quatre Bras as an eighteenth century wargames scenario can be found in:

Grant, Charles S, *The Wolfenbüttel War* (Partizan Press, 2012) (pp. 28–38).

Zone 1

SCENARIO 12: AN UNFORTUNATE OVERSIGHT

SITUATION
The Red general has been ordered to hold a strategic bridge. He has however neglected to patrol the entire river, and is therefore unaware of the existence of the ford. The Blue army is poised to exploit this oversight.

ARMY SIZES
Both armies have 6 units.

DEPLOYMENT
1. The Red army deploys first. All units are arrayed north of the river, within 12" of the town.
2. The Blue army deploys second, anywhere south of the river. All units are deployed.

REINFORCEMENTS
There are no reinforcements in this scenario.

SPECIAL RULES
1. **Scouting.** The Blue player is assumed to be engaged in patrolling. As a result, Blue units may not fire on Turn 1.

GAME LENGTH AND TURN ORDER
This scenario lasts 15 game turns. The Blue player goes first in each turn.

VICTORY CONDITIONS
Victory is secured by being in exclusive occupation of the hill at the end of the game.

INSPIRATION
This game was derived from another classic scenario from Charles Stewart Grant – although anyone who refers to the original will note some rather extensive terrain modifications. The special rule allows the otherwise docile Red army a chance to inflict casualties before the Blue forces cross the river in strength.

FURTHER READING
This scenario was adapted from one of many in the following book:

Grant, Charles Stewart, *Scenarios for Wargames* (Wargames Research Group, 1981) (pp. 16–17).

Wargame Scenarios 89

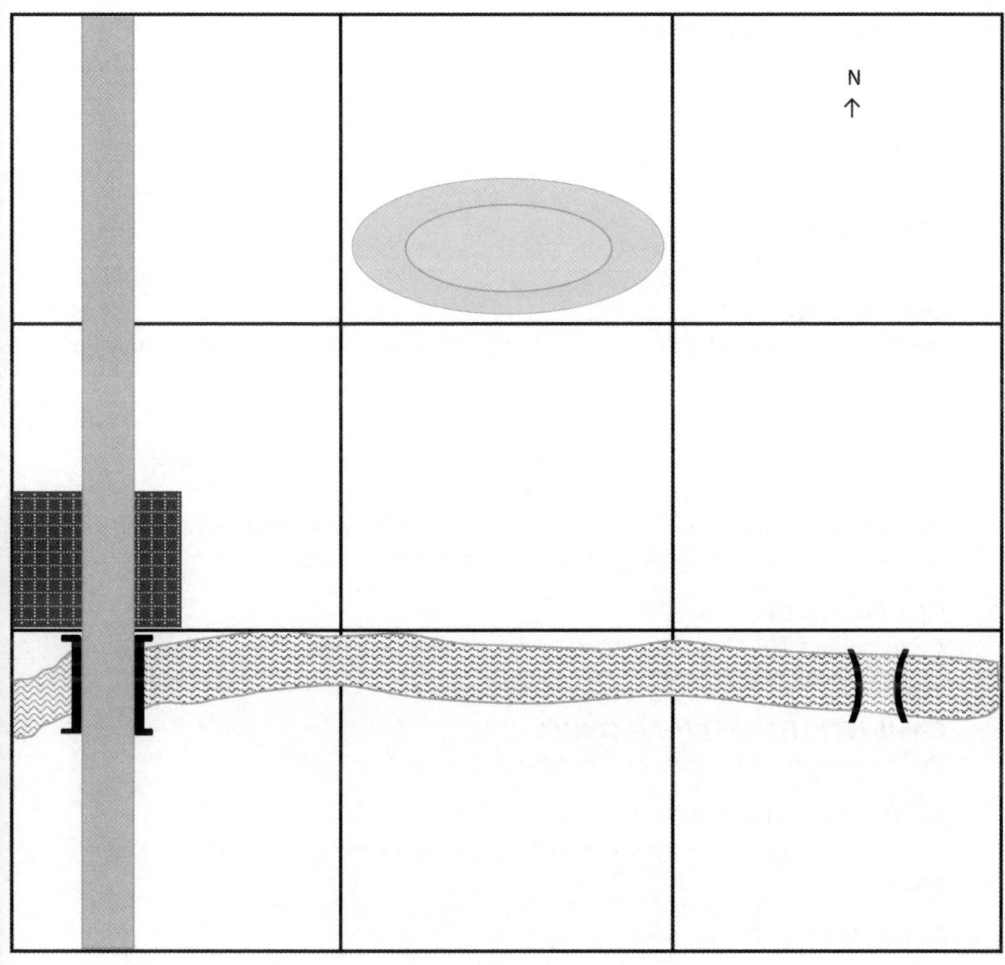

SCENARIO 13: ESCAPE

SITUATION
The Blue army is returning home after raiding Red territory. Its journey is blocked by a number of Red units.

ARMY SIZES
Both armies have 6 units.

DEPLOYMENT
The Red army deploys 1 unit in Zone 1, facing north. No Blue units are deployed at the start of this scenario.

REINFORCEMENTS
Turn 1: (a) Blue army: all 6 units appear via the road on the northern table edge.
Turn 2: (a) Red army: 2 units appear on the hill.
Turn 4: (a) Red army: 2 units arrive from the western table edge, north of the wood.
Turn 6: (a) Red army: 1 unit appears from the southern table edge.

SPECIAL RULES
1. **Exiting the Table**. Only Blue units may exit the table. They may only do so via the road on the southern table edge.

GAME LENGTH AND TURN ORDER
This scenario lasts 15 game turns. The Blue player goes first in each turn.

VICTORY CONDITIONS
The Blue player must exit 3 units from the table in order to win. Failure to do so constitutes a Red victory.

INSPIRATION
This game is another adaptation of a Charles Stewart Grant scenario. It should see the Blue player constantly on the verge of victory, only to be stymied by the appearance of more Red units.

FURTHER READING
The scenario which inspired the idea behind this game can be found in:

Grant, Charles Stewart, *Scenarios for Wargames* (Wargames Research Group, (1981) (pp. 30–31).

N
↑

Zone 1

SCENARIO 14: STATIC DEFENCE

SITUATION
The Red general has been ordered to hold both the hill and the town. His Blue counterpart has been assigned the task of capturing either the hill or the town.

ARMY SIZES
Both armies have 6 units.

DEPLOYMENT
The Red army deploys 3 units within 12" of the hill, and 3 units within 12" of the town. No Blue units are deployed at the start of this scenario.

REINFORCEMENTS
Turn 1: (a) Blue army: all 6 units appear from the southern table edge.

SPECIAL RULES
1. **Static Posture.** 2 Red units must remain within 12" of the hill, and 2 more Red units must remain within 12" of the town.

GAME LENGTH AND TURN ORDER
This scenario lasts 15 game turns. The Blue player goes first in each turn.

VICTORY CONDITIONS
The Blue player wins by being in exclusive occupation of either the hill or the town at the end of the game. Failure to do so constitutes a Red victory.

INSPIRATION
Board wargames can provide a rich source of ideas for miniature games. This scenario was inspired by one that featured in the classic boardgame *Panzer Leader*. The situation allows the attacker to seize the initiative by choosing the time and place of his or her assault.

FURTHER READING
A board wargame could scarcely be said to qualify as a book, but details are included below for readers who wish to track it down (second hand dealers will have to be the source, since the game has been unavailable for some time):

Panzer Leader (Avalon Hill Game Company, 1974) (Scenario 1, Utah Beach).

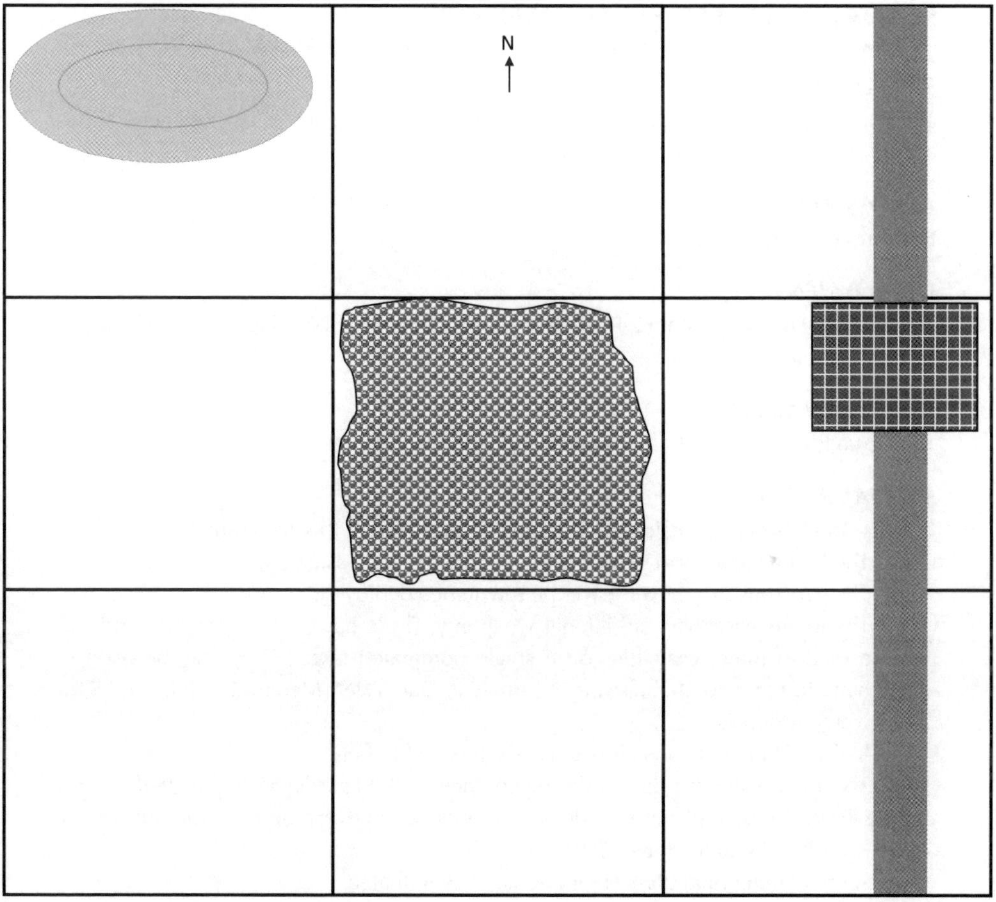

Wargame Scenarios 93

SCENARIO 15: FORTIFIED DEFENCE

SITUATION
The Red army is expecting an attack from a much larger Blue force. The Red general has accordingly prepared a fortified position for his troops.

ARMY SIZES
Both armies have 6 units.

DEPLOYMENT
The Red army deploys within 24" of the northern table edge. No Blue units are deployed at the start of this scenario.

REINFORCEMENTS
Turn 1: (a) Blue army: all 6 units appear from the southern table edge.

SPECIAL RULES
1. **Forts**. The following considerations apply to the towns in this scenario:
    (a) The Red general must deploy one unit in each town as garrisons.
    (b) The garrisons may never leave the town once deployed.
    (c) The towns each have additional weaponry. These have a range of 12", and roll a dice to inflict casualties on a single nominated target. They may be used in hand-to-hand combat if using the Ancient, Dark Ages, Medieval or Pike and Shot wargames rules.
    (d) Units within the towns always have a 360° field of fire.
    (e) The additional weaponry is destroyed once the Red garrison is eliminated.
2. **Blue Refit**. The Blue player may declare a refit once in every game, at the start of any turn. This has the following effects:
    (a) All remaining Blue units are immediately eliminated.
    (b) The entire Blue army reappears from the southern table edge as reinforcements. All units are at full strength.

GAME LENGTH AND TURN ORDER
This scenario lasts 15 game turns. The Blue player goes first in each turn.

VICTORY CONDITIONS
The Blue player wins by occupying both towns at the end of the game.

INSPIRATION
Deriving inspiration from heroism in a losing cause has long been seen as a British characteristic. One classic example of such an event was the Battle of Fontenoy (1745), when British infantry launched an assault against numerically superior French opponents, who also enjoyed the advantages of a fortified position. Amazingly, the assault nearly succeeded; the battle has inspired students of British military history ever since. This scenario has adjusted the numerical strengths radically in order to give the attacking Blue player a chance of victory.

FURTHER READING

The Battle of Fontenoy has played a great role in the history of wargaming, for both Charles Grant and his son, Charles Stewart, were so moved by the encounter that they wrote excellent books covering it. The third work cited below gives a most atmospheric account of Fontenoy; all three are however well worth reading:

Grant, Charles, *The Battle of Fontenoy* (William Luscombe, 1975).
Grant, Charles S with Phil Olley, *Wargaming in History Volume 2* (Ken Trotman, 2010) (pp. 80–123).
Rogers, Col. H.C.B; *The British Army of the Eighteenth Century* (George Allen and Unwin, 1977) (pp. 186–204).

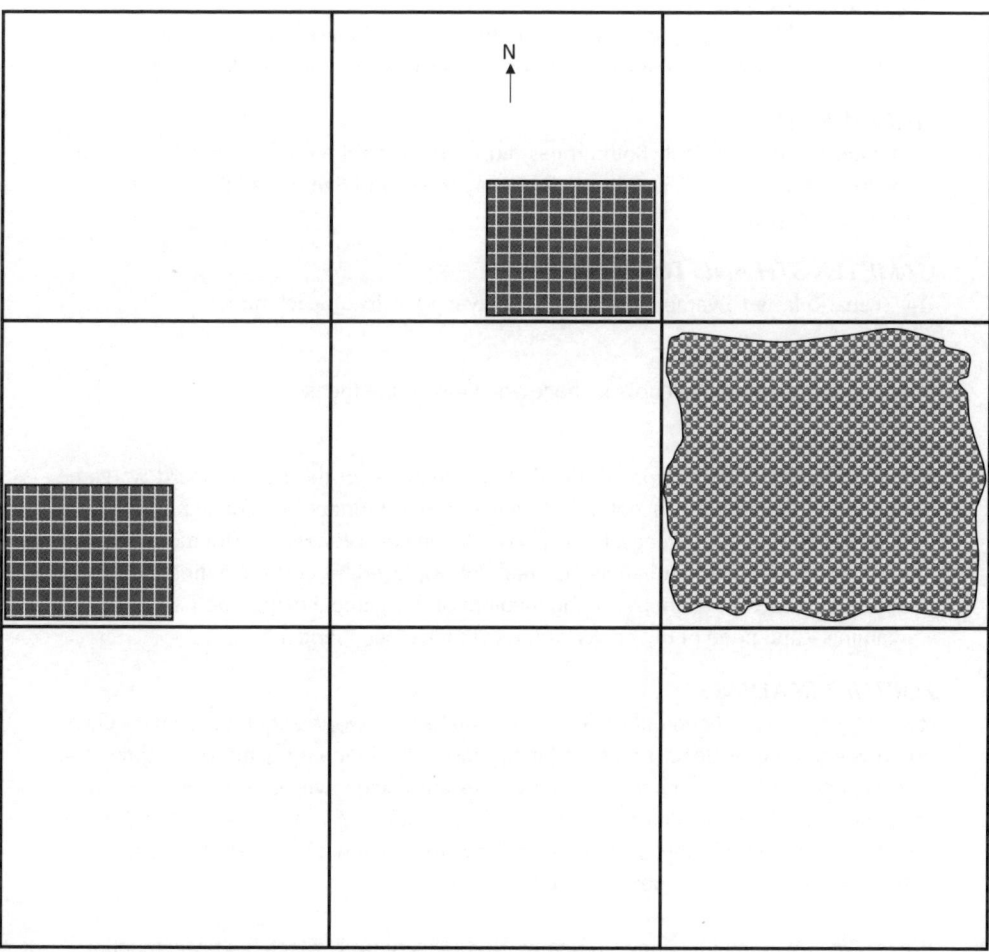

SCENARIO 16: ADVANCE GUARD

SITUATION
The Red and Blue armies represent patrols sent by larger forces to seize an outlying town. Each is unaware of the other's presence.

ARMY SIZES
Both armies have 6 units.

DEPLOYMENT
No units are deployed at the start of this game.

REINFORCEMENTS
Turn 1: (a) Red army: all 6 units arrive from the road on the northern table edge.
(b) Blue army: all 6 units arrive from the road on the southern table edge.

SPECIAL RULES
1. **Blundering Into Contact**. Both armies must remain on the road, proceeding towards the town at a rate of 9" per move. They may move and fight normally as soon as the town is occupied.

GAME LENGTH AND TURN ORDER
This scenario lasts 15 game turns. The Red player goes first in each turn.

VICTORY CONDITIONS
Victory goes to the side occupying the town at the end of the game.

INSPIRATION
This scenario is an adaptation of one that was featured in the classic board wargame *Panzer Blitz*. The central concept of having two armies blunder into contact represents a particularly intriguing challenge; but I also wanted to cover *Panzer Blitz* for more personal reasons. For if my cousin, Julian Stokes, had not displayed his customary fine judgement in introducing me to the hobby via the medium of this game, I may never have taken up wargaming – and none of my books would ever have been written.

FURTHER READING
Panzer Blitz has long been out of production, but is well worth acquiring from specialist board wargaming dealers. It was by far the most revolutionary game of its time, and its design genius was – rather gratifyingly – instantly acknowledged by the wargaming community. Its designer, James F. Dunnigan, is generally regarded as the most influential board wargamer of all time, and has since become a highly respected commentator on military affairs. The game's details are cited below:

Panzer Blitz (Avalon Hill Game Company, 1970) (Scenario 7, Meeting Engagement).

Wargame Scenarios 97

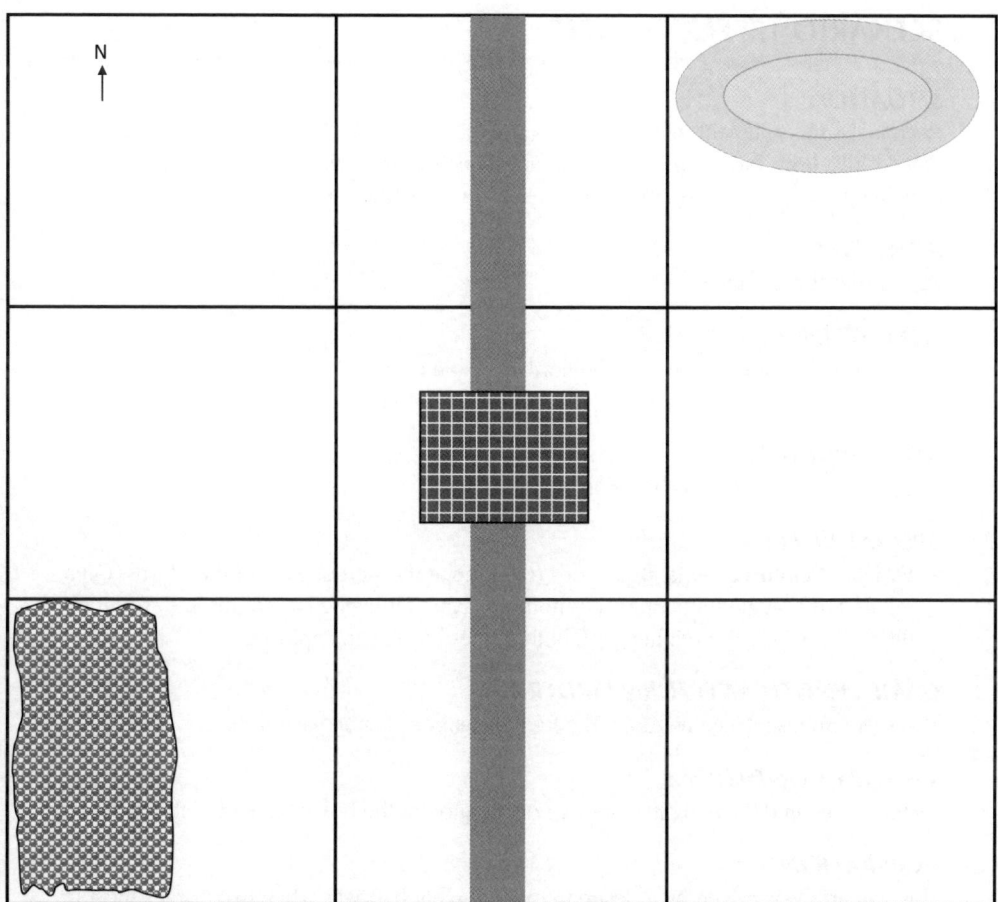

SCENARIO 17: ENCOUNTER

SITUATION
Reconnaissance elements of both the Red and the Blue armies have located a strategically placed hill. Both have sent word to their respective generals, who have ordered their remaining units to arrive on the scene as soon as possible.

ARMY SIZES
Both armies have 6 units.

DEPLOYMENT
1. Red army: 1 unit within 6" of the northern table edge.
2. Blue army: 1 unit within 6" of the southern table edge.

REINFORCEMENTS
See special rule for Variable Reinforcements below.

SPECIAL RULES
1. **Variable Reinforcements**. Both sides roll a die at the start of each of their turns. On a roll of 4–6 a single unit appears from the relevant table edge (north for Red, south for Blue). This process continues until both armies have been deployed.

GAME LENGTH AND TURN ORDER
This scenario lasts 15 game turns. The Red player goes first in each turn.

VICTORY CONDITIONS
Victory is secured by being in exclusive occupation of the hill at the end of the game.

INSPIRATION
This scenario was driven by a desire to explore what happens when two armies meet in a haphazard manner. Much will depend upon luck (the arrival of reinforcements); great skill will be required to overcome the vicissitudes of ill-fortune – this represents a far more interesting challenge than a totally balanced encounter.

FURTHER READING
This game is loosely based upon another excellent scenario featured in the book listed below:

Grant, Charles Stewart, *Scenarios for Wargames* (Wargames Research Group, 1981) (pp. 97–98).

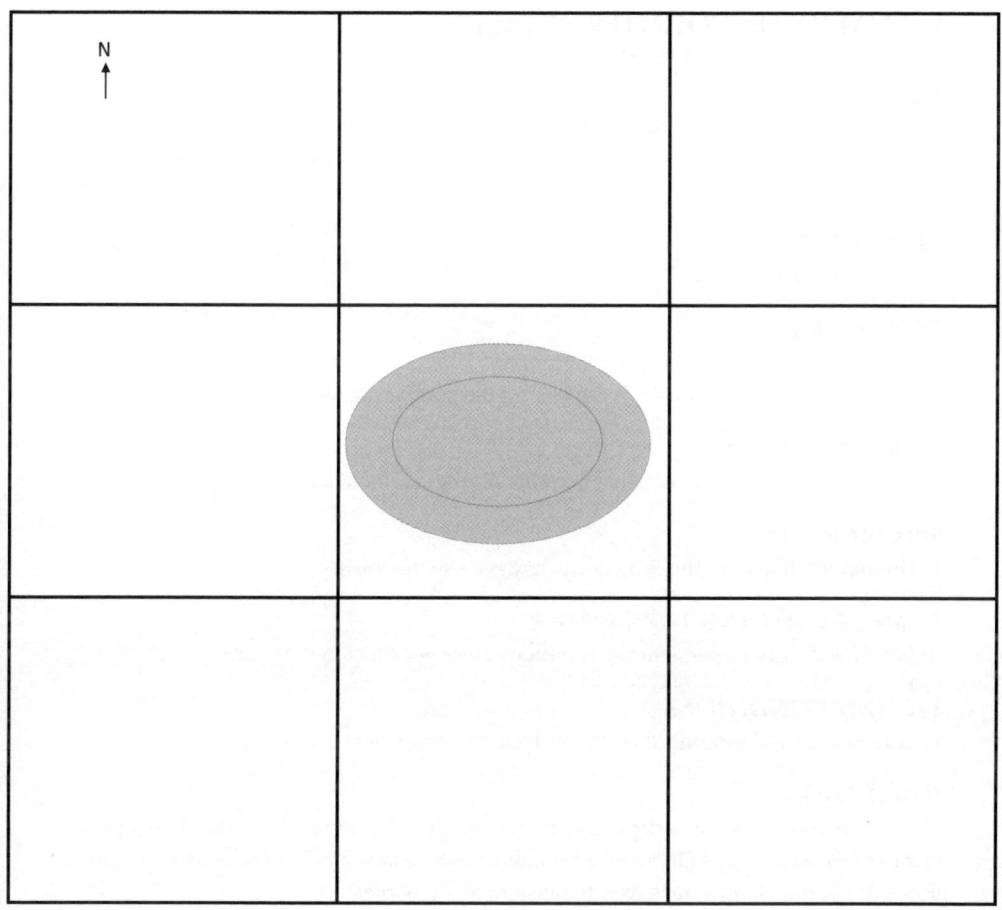

SCENARIO 18: COUNTER-ATTACK

SITUATION
The Blue army has been ordered to seize the bridge. The Blue general has not located the fords or (more importantly) the Red army, whose commander is preparing a counter-attack.

ARMY SIZES
Both armies have 6 units.

DEPLOYMENT
1. Red army: 1 unit in Zone 1, facing south.
2. Blue army: all 6 units deploy within 6" of the southern table edge.

REINFORCEMENTS
Turn 3: (a) Red army: 5 units arrive from the northern table edge.

SPECIAL RULES
1. **Limited Intelligence**. The Blue army may not use the fords.

GAME LENGTH AND TURN ORDER
This scenario lasts 15 game turns. The Blue player goes first in each turn.

VICTORY CONDITIONS
Victory is achieved by control of the bridge and occupation of the town.

INSPIRATION
This game represents an adaptation of the Battle of Langensalza (1866), when the Hanoverian (Red) forces launched a counter-attack against a rather complacent Prussian (Blue) army. The Hanoverians won the historical encounter.

FURTHER READING
Accounts of the Battle of Langensalza can be found in:

Barry, Quintin, *The Road to Königgrätz* (Helion, 2010) (pp. 208–222).
Pocock, John, *Langensalza 1866* (Continental Wars Society, 2002).

John Pocock's booklet cited above contains much helpful advice on wargaming the battle. An excellent wargame scenario covering Langensalza can be found in:

Weigle, Bruce, *1866* (Medieval Miscellanea, 2010) (pp. 47–48).

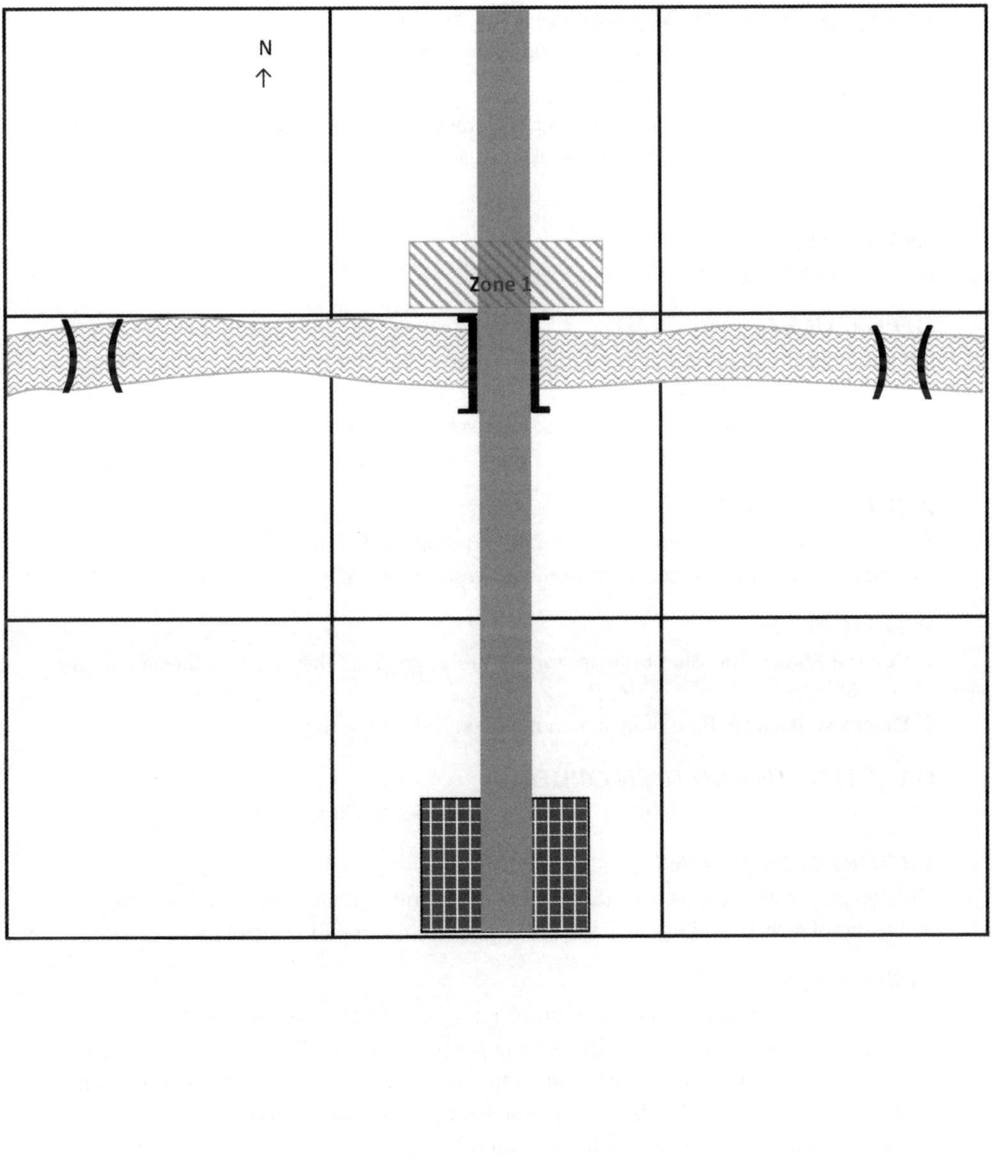

Wargame Scenarios

SCENARIO 19: BLOW FROM THE REAR

SITUATION
The Blue general is defending two river crossings from what he thinks is a numerically inferior Red force. His complacency will be shattered when an additional Red contingent arrives in the Blue rear area.

ARMY SIZES
Both armies have 6 units.

DEPLOYMENT
No Red units are deployed at the start of the game. The Blue army is arrayed as follows:

(a) 4 units south of the river within 6" of the riverbank, facing north.
(b) 2 units in zone 1, facing north.

REINFORCEMENTS
Turn 1: (a) Red army: 3 units arrive from the northern table edge.
Turn 6: (a) Red army: 3 units arrive from the western table edge, south of the river.

SPECIAL RULES
1. **Reserve Status**. The Blue units in zone 1 are in reserve. They may neither move nor fire until turn 7.
2. **Defensive Posture**. Blue units may not move north of the river.

GAME LENGTH AND TURN ORDER
This scenario lasts 15 game turns. The Red player goes first in each turn.

VICTORY CONDITIONS
The Red player wins the game if there are no Blue units within 6" of either river crossing at the end of turn 15. Failure to achieve this goal results in a Blue victory.

INSPIRATION
This scenario is based upon the Battle of Krefeld (1758), which saw Ferdinand of Brunswick's Hanoverian army (Red) rout a larger but indifferently-led French force under the Count of Clermont. The superior quality and leadership of Ferdinand's army is not covered in the wargame, but is instead depicted by an equalization of forces. The French lack of preparation is reflected by the special rules.

FURTHER READING
The following book not only provides a fine account of the battle, but also depicts two outstanding wargaming re-enactments of it:

Grant, Charles S, with Phil Olley, *Wargaming in History Volume 1* (Ken Trotman, 2009) (pp. 33–64).

Wargame Scenarios 103

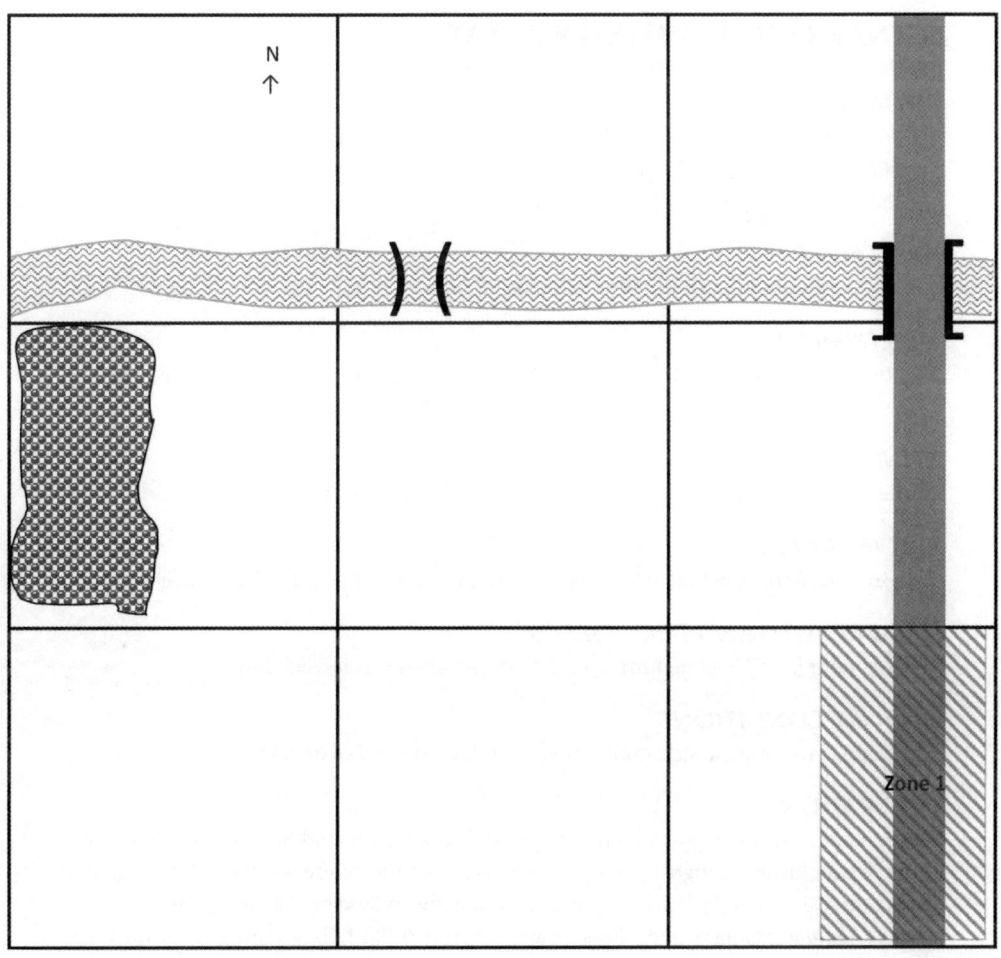

SCENARIO 20: FIGHTING RETREAT

SITUATION
The Red army has been raiding Blue territory. The army of the latter is in hot pursuit of the raiders, who are equally determined to make their escape by crossing the river, and consolidating their position on a dominant hill.

ARMY SIZES
The Red army has 4 units; the Blue army has 6 units.

DEPLOYMENT
The Red army deploys all 4 units anywhere south of the river. No Blue units are deployed at the start of the game.

REINFORCEMENTS
Turn 2: (a) Blue army: 6 units arrive from the southern table edge.

SPECIAL RULES
1. **Retreat Orders**. Any Red army unit south of the river at the end of Turn 2 is eliminated.

GAME LENGTH AND TURN ORDER
This scenario lasts 15 game turns. The Red player goes first in each turn.

VICTORY CONDITIONS
The game is won by the side controlling the hill at the end of turn 15.

INSPIRATION
This scenario saw its origins in a purchase made on a Bring and Buy sale at a wargames show. I was lucky enough to find several issues of the renowned (and long defunct) magazine *Wargamer's Newsletter*, which under the editorship of the great wargames pioneer Donald Featherstone, did so much to spread the hobby during the 1960s. The article upon which I have drawn is an absolute classic; it contains a map, full scenario details, a wargames battle report, and a fine set of simple rules for the medieval period. It set an example of thoroughness, precision and concision that few modern wargames writers are able to emulate.

My own adaptation reproduces the core concept of the fighting retreat, but has made considerable alterations in order to squeeze the game onto a small table.

FURTHER READING
The original scenario can be found in:

Wargamer's Newsletter no. 69 (December 1967) 'Cry God for England, Harry and Saint George!' by Don Featherstone.

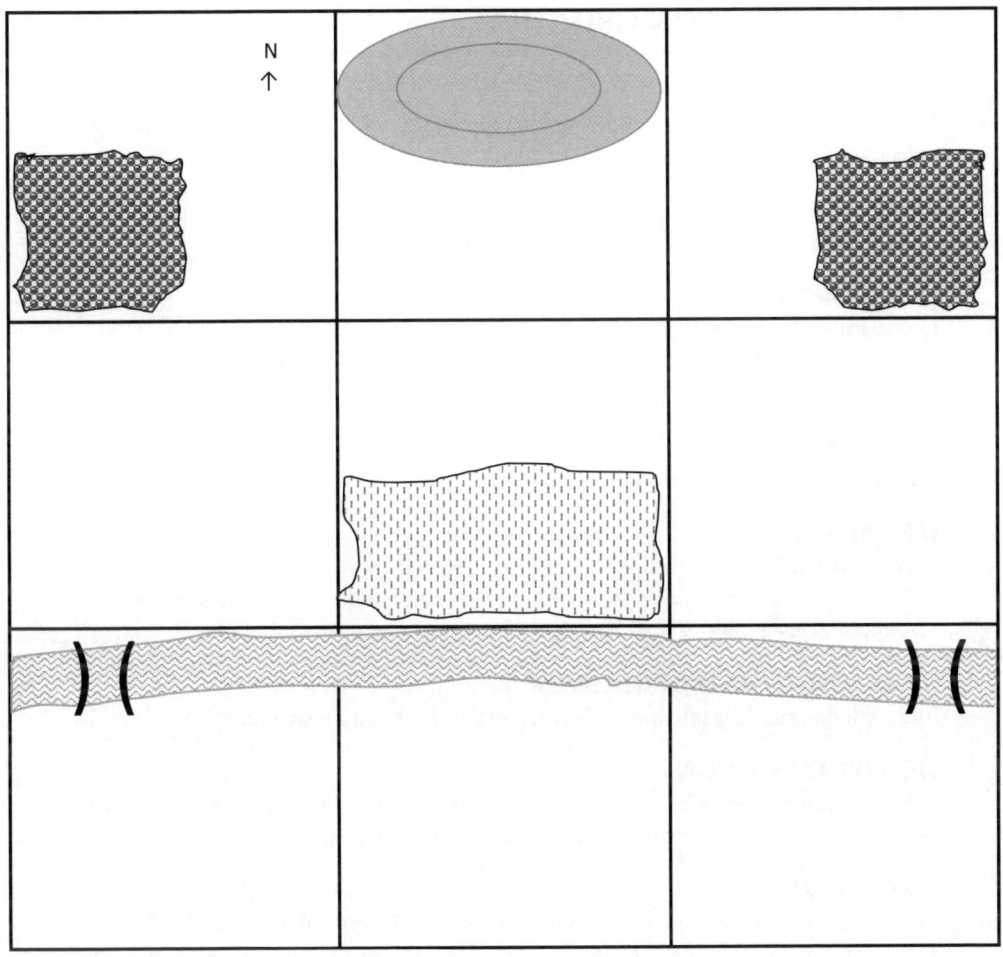

SCENARIO 21: TWIN OBJECTIVES

SITUATION
The Red army is split between holding the town and the hill. The Blue general has been ordered to take both objectives.

ARMY SIZES
The Red army has 4 units; the Blue army has 6 units.

DEPLOYMENT
1. Red army: (a) 1 unit on the hill, facing east.
    (b) 3 units within 6" of the northern table edge, facing south.
2. Blue army: all 6 units in zone 1, facing north.

REINFORCEMENTS
There are no reinforcements in this scenario.

SPECIAL RULES
1. **Wooded Hill**. The hill is slightly wooded (denoted by placing a few trees upon it). This makes it impassable to all troops except Infantry, Warband, Levies, Swordsmen, Archers, Zouaves and Skirmishers. It also confers cover against shooting.

GAME LENGTH AND TURN ORDER
This scenario lasts 15 game turns. The Blue player goes first in each turn.

VICTORY CONDITIONS
The Blue player wins by being in exclusive occupation of both the hill and the town at the end of the game. Failure to do so constitutes a Red victory.

INSPIRATION
This scenario derives from the Battle of Lobositz (1756), which saw the Prussians (Blue army) defeat the Austrians (Red army). The division of effort provides for a most interesting game: this was appreciated by the Grant family, three generations of whom (the late Charles, his son Charles Stewart, and grandson Charles Murray), have taken part in wargames covering the battle.

FURTHER READING
The first work, a biography of the Austrian commander Field Marshal von Browne, gives an excellent historical account and analysis of the Battle of Lobositz; the second provides details on how to wargame the encounter:

Duffy, Christopher, *The Wild Goose and the Eagle* (Tricorne, 2009; originally 1964) (pp. 210–222).
Grant, C. S. *The War Game Companion* (Ken Trotman, 2008) (pp. 104–125).

Wargame Scenarios 107

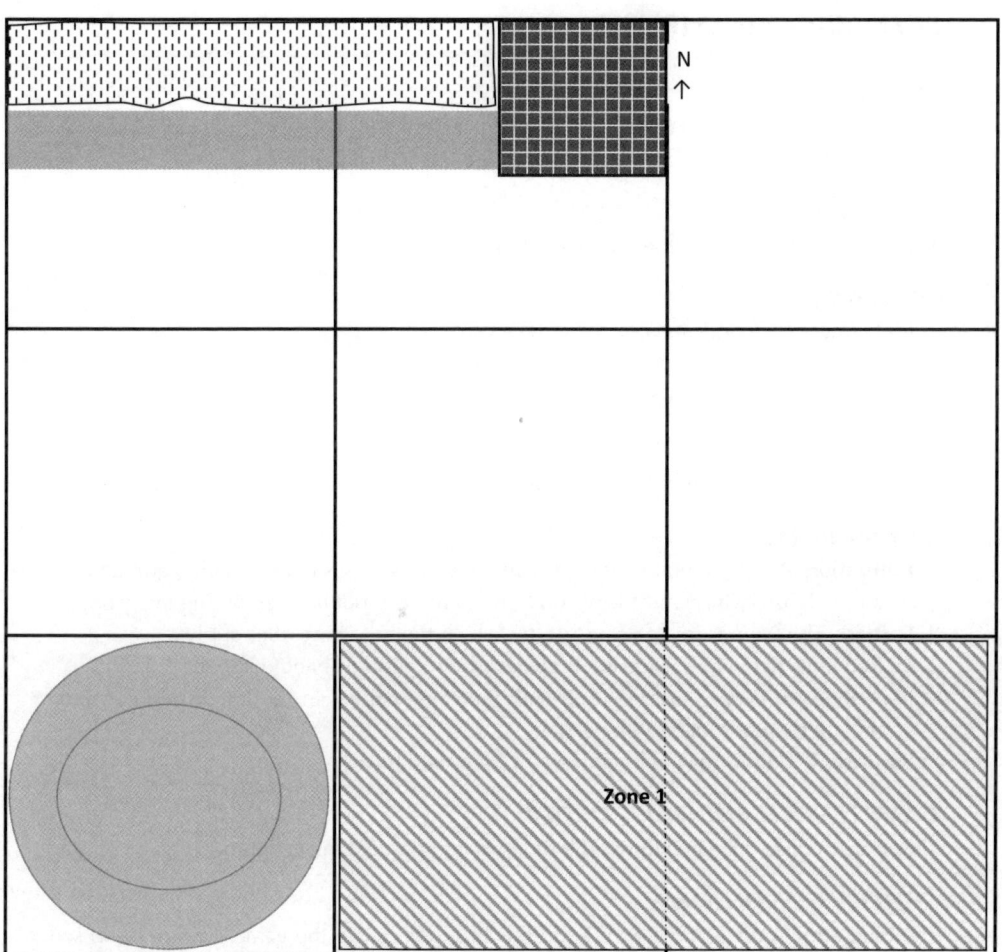

SCENARIO 22: AMBUSH

SITUATION
The Red army is besieging a Blue fort. The Red general has however neglected to provide adequate sentries, allowing a small Blue force to prepare an ambush.

ARMY SIZES
The Red army has 6 units; the Blue army has 4 units.

DEPLOYMENT
1. Red army: all 6 units in zone 1, facing west.
2. Blue army: (a) 1 unit in the town, facing east.
   (b) 3 units in the woods, facing north.

REINFORCEMENTS
There are no reinforcements in this scenario.

SPECIAL RULES
1. **Infiltration**. Any Blue unit may deploy in the woods. Units not normally permitted to do so (such as Cavalry) must leave on turn 1, and may not re-enter during the game.
2. **Fortress**. The town is treated as a fort, for which the following rules apply:
   (a) Red units may not engage the Blue garrison in hand-to-hand combat.
   (b) A Blue Artillery, Mortar or Anti-tank unit may deploy in the fort if desired, and remain there throughout the game.
3. **Surprise**. Red units may neither move nor fire until turn 3.

GAME LENGTH AND TURN ORDER
This scenario lasts 15 game turns. The Blue player goes first in each turn.

VICTORY CONDITIONS
The Blue army must eliminate all Red units in order to win the game. Failure to do so constitutes a Red victory.

INSPIRATION
An ambush is a classic wargame scenario, allowing as it does for a numerically inferior force to turn the tables against its foe. The Battle of Auberoche (1345), which saw the English (Blue) defeat the French (Red) army is a particularly fine example.

FURTHER READING
Donald Featherstone has penned a customarily outstanding account both of the battle, and its re-creation as a wargame scenario, in:

Featherstone, Donald F, *Wargaming: Ancient and Medieval* (David & Charles, 1975) (pp. 95–101).

Wargame Scenarios 109

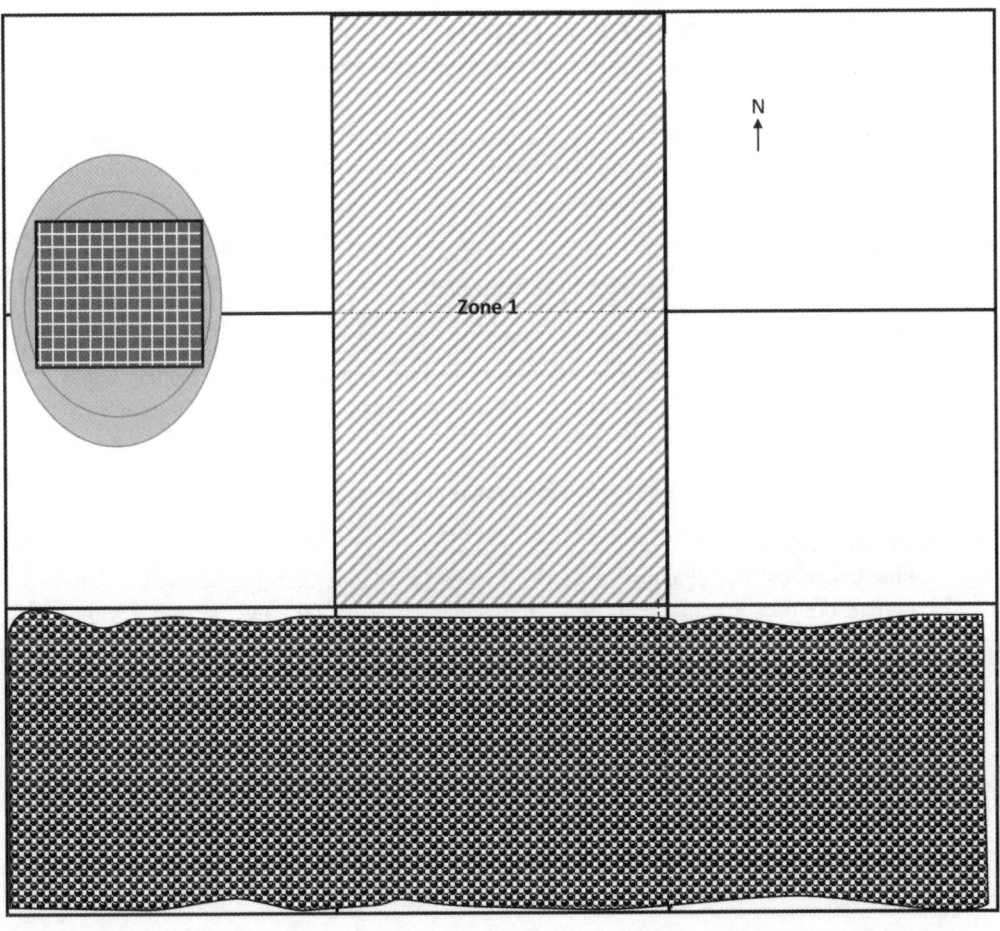

SCENARIO 23: DEFENCE IN DEPTH

SITUATION
The Red army has been suppressing an insurrection. It is marching towards the rebel (Blue) headquarters. However, the Blue forces, supported by some local irregular troops, are preparing an unpleasant surprise.

ARMY SIZES
The Red army has 6 units; the Blue army has 4 units.

DEPLOYMENT
The Blue army may place 1 unit in the woods north of the river if desired; the remaining 3 units are placed anywhere south of the river. No Red units are deployed at the start of the game.

REINFORCEMENTS
Turn 1: (a) Red army: all 6 units arrive from the northern table edge.

SPECIAL RULES
1. **Blue Irregulars**. The Blue army includes 2 Irregular infantry units (replace any 2 other units as desired). Irregular units have a movement distance of 9", and may move within woods; they are equipped with missile weapons which have a range of 12"; and they reduce all combat die rolls by 2.
2. **Unlimited Ammunition**. (This rule only applies to Pike and Shot wargames). Blue Irregular units are equipped with bows; they never run out of ammunition.
3. **Lack of Local Knowledge**. Red units may not enter woods.
4. **Exiting the Table**. Red units may exit the southern table edge via the road.

GAME LENGTH AND TURN ORDER
This scenario lasts 15 game turns. The Red player goes first in each turn.

VICTORY CONDITIONS
The Red player must exit 3 units from the table in order to win. Failure constitutes a Blue victory.

INSPIRATION
Insurgent warfare offers scope for a slightly different but extremely challenging wargame, allowing as it does for inferior forces to take advantage of difficult terrain, thereby springing a surprise upon an unprepared foe. This game is loosely based upon the Battle of the Yellow Ford (1598), fought between English (Red) and Irish (Blue) armies; it also derives some inspiration from one of Charles Stewart Grant's scenarios.

FURTHER READING
The core ideas behind this wargame can be found in:

Grant, Charles Stewart, *Scenarios for Wargames* (Wargames Research Group, 1981) (pp. 32–33).
Wesencraft, C.F; *With Pike and Musket* (Elmfield Press, 1975) (pp. 82–85).

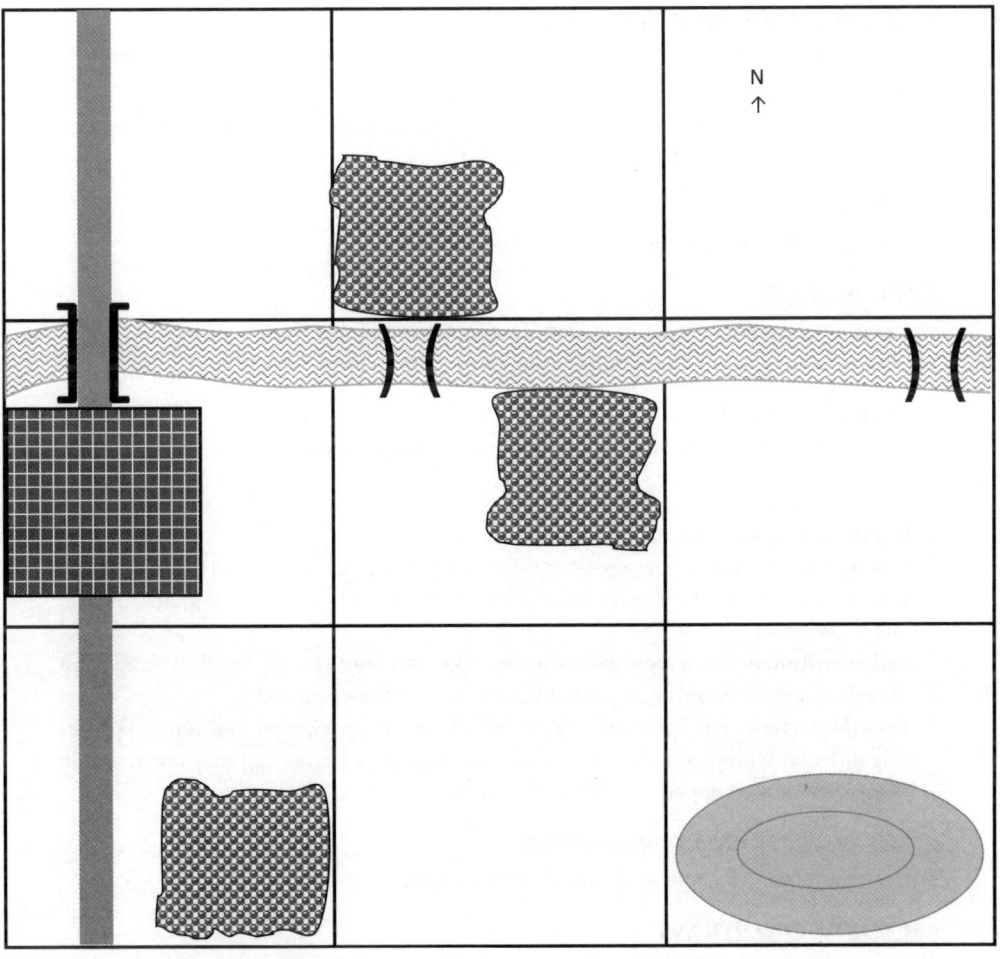

SCENARIO 24: BOTTLENECK

SITUATION
The Blue army outnumbers its opponents, but its attempts to clear the road are hampered by a lake and an apparently impenetrable forest.

ARMY SIZES
The Red army has 4 units; the Blue army has 6 units.

DEPLOYMENT
The Red army deploys 1 unit in the wood, and the remaining 3 units within 12" of the northern table edge. No Blue units are deployed at the start of this game.

REINFORCEMENTS
Turn 1: (a) Blue army: all 6 units arrive from the southern table edge.

SPECIAL RULES
1. **Forest.** Blue army units may not enter the wood.
2. **Red Skirmishers.** For wargames set in the Ancient, Dark Ages, Rifle and Sabre or Horse and Musket periods, the Red army must include 1 unit of Skirmishers. Replace another unit if necessary.
3. **Red Swordsmen.** For wargames set in the Pike and Shot period, the Red army must include 1 unit of Swordsmen. Replace another unit if necessary.
4. **Peasant Archers.** For wargames set in the Medieval period, the Red army replaces any one unit with peasant archers. These are treated as Levies, but may move within woods and are equipped with bows (12" range).

GAME LENGTH AND TURN ORDER
This scenario lasts 15 game turns. The Blue player goes first in each turn.

VICTORY CONDITIONS
The Blue player wins if no Red units in open terrain are within 6" of the road at the end of the game (units in the wood do not count). Failure to achieve this condition results in a Red victory.

INSPIRATION
Bottlenecks have often allowed armies of inferior strength to hold a position against superior foes. This game represents an attempt to simulate this rather tricky conundrum; it is based upon the Battle of Thames River (1813) fought during the War of 1812 between the British (Red) and American (Blue) armies.

FURTHER READING
The following outstanding book by Stuart Asquith provides a veritable treasure trove of wargames, and will open any gamer's eyes to the largely neglected War of 1812, in the same way that it inspired this scenario:

Asquith, Stuart, *Scenarios for the War of 1812* (Partizan Press, 2010) (pp. 51–54).

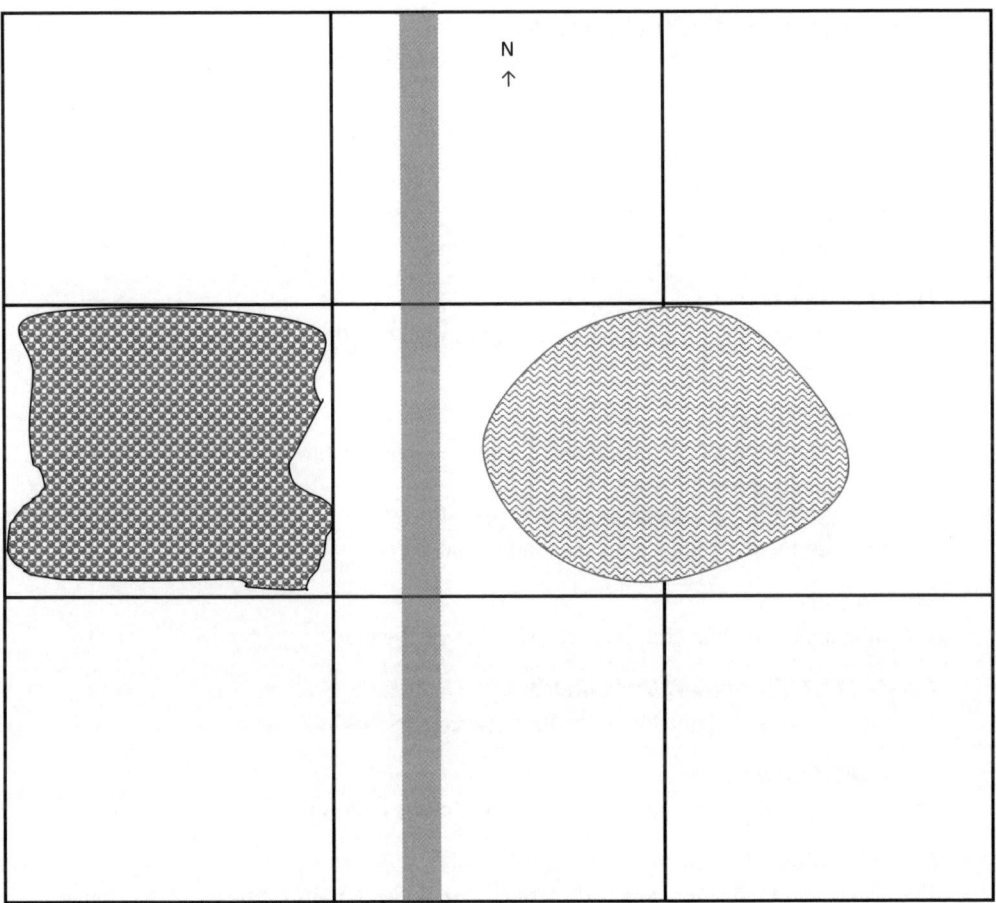

114   One-Hour Wargames

SCENARIO 25: INFILTRATION

SITUATION
A small Blue force has penetrated the Red lines. It aims to plunder the Red army's supply network before the latter can react.

ARMY SIZES
The Red army has 6 units; the Blue army has 4 units.

DEPLOYMENT
The Red army deploys 1 unit in zone 1, facing south. No Blue units are deployed at the start of the game.

REINFORCEMENTS
Turn 1: (a) Blue army: all 4 units arrive from the southern table edge, within 12" of the south-western corner.
Turn 3: (a) Red army: 2 units from the northern table edge.
Turn 6: (a) Red army: 3 units from the southern table edge, via the road.

SPECIAL RULES
1. **Exiting the Table**. Blue units may exit the table via the road on the northern table edge.

GAME LENGTH AND TURN ORDER
This scenario lasts 15 game turns. The Blue player goes first in each turn.

VICTORY CONDITIONS
The Blue player must exit 2 units from the table. Failure to do so constitutes a Red victory.

INSPIRATION
This game retains the numerical disparities and some terrain elements of the Battle of Kernstown (1862), but has converted what was a pitched battle into an infiltration scenario. This allows an otherwise one-sided affair to become rather more interesting.

FURTHER READING
Much of my youth was spent avidly reading *Strategy and Tactics* magazine, which featured (and still does) a board wargame in every issue. The article and boardgame map on Kernstown proved especially helpful; I have also cited a more readily available account:

Dougherty, Kevin et al; *Battles of the American Civil War 1861–1865* (Amber, 2007) (pp. 50–59).
Nofi, Al, 'Stonewall in the Shenandoah' in *Strategy and Tactics No. 67* (March, 1978) (pp. 4–13).

Wargame Scenarios 115

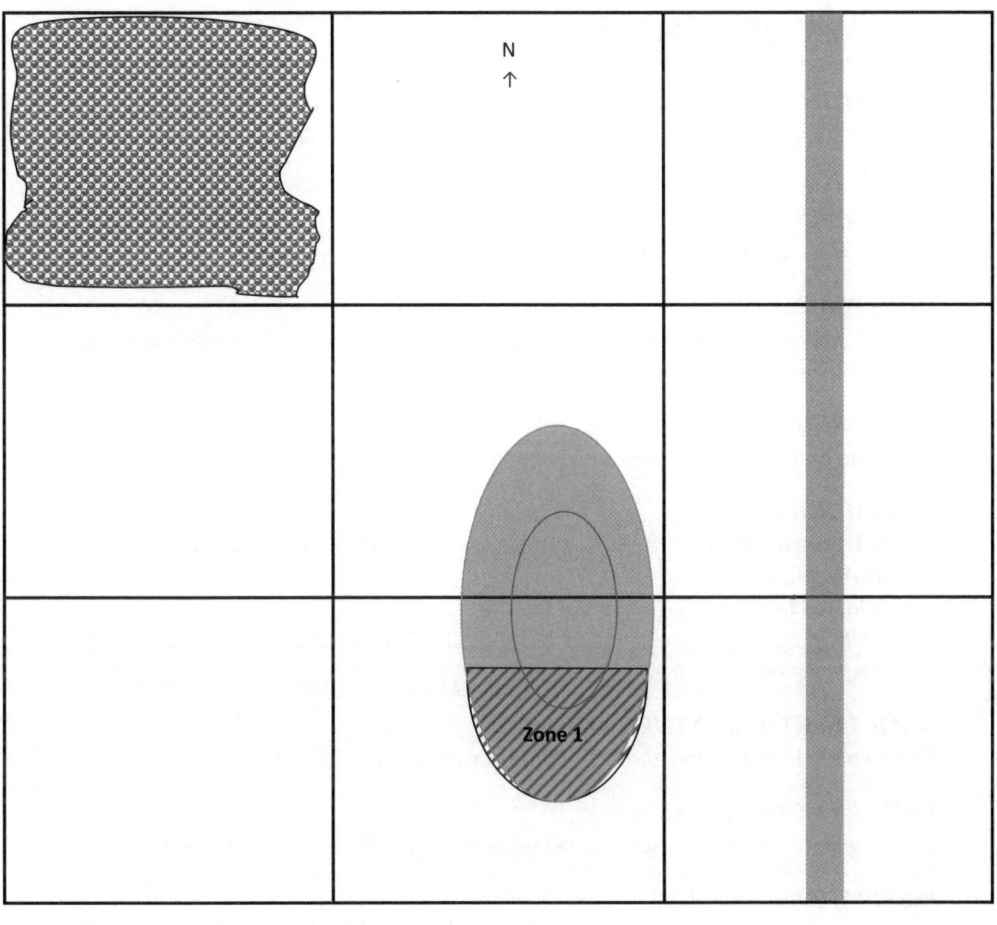

SCENARIO 26: TRIPLE LINE

SITUATION
The Blue army has invaded Red territory. A rather ad hoc and disorganized collection of Red units has assembled to meet the threat.

ARMY SIZES
The Red army has 4 units; the Blue army has 6 units.

DEPLOYMENT
The Red army has 1 unit in zone 1, facing south; 2 units in zone 2, facing south; and 1 unit on the hill facing south. No Blue units are deployed at the start of the game.

REINFORCEMENTS
Turn 1: (a) Blue army: all 6 units arrive from the southern table edge.

SPECIAL RULES
1. **Red Disorganization**. The disarray engendered by the rapid Blue assault has the following effects:
   (a) No Red unit may move south of the river.
   (b) Individual Red units may never move until a Blue unit moves within 6". Firing is unrestricted.

GAME LENGTH AND TURN ORDER
This scenario lasts 15 game turns. The Blue player goes first in each turn.

VICTORY CONDITIONS
Victory goes to the side in exclusive occupation of the hill, at the end of the game.

INSPIRATION
This game stems from another of Stuart Asquith's splendid War of 1812 scenarios. The terrain has been changed slightly; the confusion of the American (Red) forces has been reflected but their numbers have not. The real Battle of Bladensburg (1814) saw the Americans enjoying substantial numerical superiority but appalling moral deficiencies (many units ran away at the first shot). I have accordingly allowed the British to enjoy the larger numbers on this occasion; the deliberate lack of specific morale provisions in the rulesets printed in this book explains my decision.

FURTHER READING
The scenario which inspired this game can be found in:

Asquith, Stuart, *Scenarios for the War of 1812* (Partizan Press, 2010) (pp. 72–76).

Wargame Scenarios 117

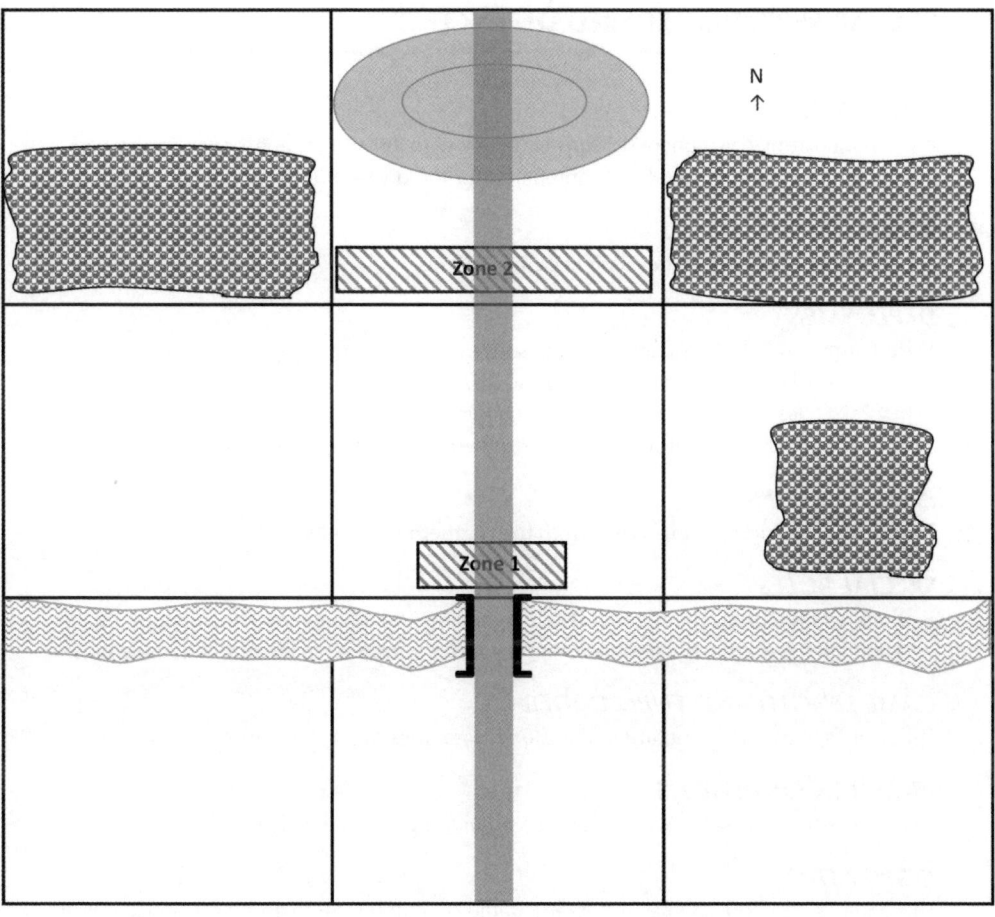

SCENARIO 27: DISORDERED DEFENCE

SITUATION
A small Blue army has launched a surprise attack with the aim of capturing the crossroads. The larger Red army is singularly unprepared for the onslaught.

ARMY SIZES
The Red army has 6 units; the Blue army has 4 units.

DEPLOYMENT
1. Red army: (a) 1 unit in zone 1, facing south.
   (b) 1 unit in zone 2, facing south.
   (c) 1 unit at the crossroads, facing south.
2. Blue army: all 4 units within 6" of the southern table edge, facing north.

REINFORCEMENTS
Turn 8: (a) Red army: 3 units arrive from the northern table edge.

SPECIAL RULES
1. **Blue Consolidation**. The Blue army is assigned the task of capturing and controlling the crossroads. Accordingly, no Blue unit may move within 4" of the northern table edge.

GAME LENGTH AND TURN ORDER
This scenario lasts 15 game turns. The Blue player goes first in each turn.

VICTORY CONDITIONS
The side occupying the crossroads at the end of the game is the victor.

INSPIRATION
This game is very loosely inspired by the Battle of Shiloh (1862), an American Civil War engagement that saw the Confederate (Blue) army launch a spectacular but eventually unsuccessful attack upon the Union (Red) forces. The deployment and reinforcement schedule bears a good deal of relationship to the game's source, but the terrain does not; the real battlefield of Shiloh was heavily wooded, which does in every sense provide an obstacle to its recreation in miniature. Also, no attempt has been made to depict the River Tennessee, from where the Union reinforcements arrived: the small size of this wargames battlefield makes its inclusion impossible.

FURTHER READING
The old magazine article listed below gives an outstanding account of Shiloh; the recently published book provides a readily accessible reference:

Berg, Richard, 'Bloody April: the Battle of Shiloh, 1862' in *Strategy and Tactics No. 76* (September, 1979) (pp. 23–30).

Dougherty, Kevin et al; *Battles of the American Civil War 1861–1865* (Amber, 2007) (pp. 60–69).

Zone 1   Zone 2

N ↑

SCENARIO 28: BOTCHED RELIEF

SITUATION
The Blue army is assaulting a town held by Red units. A Red relief force is deployed on the hill and poised to assist, but its commander is proving rather inert.

ARMY SIZES
The Red army has 6 units; the Blue army has 4 units.

DEPLOYMENT
1. Red army: (a) 2 units within 12" of the northern table edge, facing south.
   (b) 4 units on the hill, facing east.
2. Blue army: all 4 units within 6" of the southern table edge, facing north.

REINFORCEMENTS
There are no reinforcements in this scenario.

SPECIAL RULES
1. **Red Relief Force**. The units deployed on the hill may only be activated sequentially. This is realized by allowing the Red player to nominate one of these units at the start of the game; this is the only unit on the hill that may move and/or fight. Once it is eliminated, the Red player may activate another unit based on the hilltop, and so on until the final unit is activated.
2. **Blue Single Mindedness**. The Blue army has been ordered to seize the town, rather than destroy the Red force per se. As a result, Blue units may never occupy the hill; neither may they engage inactive Red units in combat.

GAME LENGTH AND TURN ORDER
This scenario lasts 15 game turns. The Blue player goes first in each turn.

VICTORY CONDITIONS
Victory is secured by occupation of the town at the game's end.

INSPIRATION
This scenario draws its inspiration from the fascinating but rather neglected Mexican-American War (1846–1848), which saw a small but extremely competent US army defeat a large yet disgracefully-led foe. The Battle of El Molino del Rey (1847) is a typical encounter, which saw the Americans prevail in a tough engagement: Mexican failure was largely due to the total inertia displayed by their relief force, supposedly poised to attack the American left flank. The situation is reflected in this wargame; there is no particular need to depict the units on the hill (they could simply be deployed off the table and arrive sequentially if desired), save to engender feelings of frustration in the mind of the Red player: he or she has units available, but may not use them as desired.

FURTHER READING
The following book gives a splendid account, not only of the Mexican-American War, itself, but also of the capabilities of the respective armies and data on their uniforms (including some colour plates). All this information makes the work an ideal source for wargamers:

Adams, Anton, *The War in Mexico* (Emperor's Press, 1998) (pp. 105–109).

Wargame Scenarios 121

SCENARIO 29: SHAMBOLIC COMMAND

SITUATION
The Blue army has been ordered to defend a strategic hill. Its superior numbers would normally make this a simple task, were it not hamstrung by appalling leadership.

ARMY SIZES
The Red army has 4 units; the Blue army has 6 units.

DEPLOYMENT
The Blue army deploys 4 units on the hill, facing north; and 2 units in zone 1, also facing north. No Red units are deployed at the start of the game.

REINFORCEMENTS
Turn 1: Red army: all 4 units arrive from the northern table edge.

SPECIAL RULES
1. **Dreadful Leadership**. Only 2 Blue units may move and/or engage in combat during each turn. The Blue player may choose which units are active.

GAME LENGTH AND TURN ORDER
This scenario lasts 15 game turns. The Red player goes first in each turn.

VICTORY CONDITIONS
The side in exclusive occupation of the hill at the game's end is victorious.

INSPIRATION
There are strong similarities between this encounter and the previous scenario, since both games see the numerically superior army suffer from terrible leadership – a situation likely to engender much frustration for the wargamer! This game is based upon the Battle of Medina de Rio Seco (1808), a Napoleonic engagement that saw the Spanish (Blue) army defeated by the French (Red) force.

FURTHER READING
Accounts of the Battle of Medina de Rio Seco can be found in:

Chandler, David (ed), *Napoleon's Marshals* (Weidenfeld & Nicolson, 1987) (pp. 70–75).
Lipscombe, Col. Nick, *The Peninsular War Atlas* (Osprey, 2010) (pp. 44–47).

Zone 1

SCENARIO 30: LAST STAND

SITUATION
The Red army has just been routed. A gallant remnant of the defeated Red forces has however survived, and is preparing to fight to the last man, in order to allow the rest of the army time to regroup.

ARMY SIZES
The Red army has 3 units; the Blue army has 6 units.

DEPLOYMENT
The Red army deploys all its units north of the river. No Blue units are deployed at the start of the game.

REINFORCEMENTS
Turn 1: Blue army: all 6 units arrive from the southern table edge.

SPECIAL RULES
1. **Elite Defenders**. Red units apply an additional modifier of +2 to all combat die rolls.
2. **Redoubt**. A single Red unit (apart from Knights Cavalry, Reiters, or Tanks) deployed on the hilltop is assumed to be in a redoubt. This has the following attributes:
   (a) The garrison is under cover.
   (b) The garrison has a field of fire of 180°.
   (c) It takes a complete turn for enemy units to enter the redoubt.
3. **Unlimited Numbers**. Any eliminated Blue units automatically reappear on the next Blue turn. They arrive from the southern table edge.

GAME LENGTH AND TURN ORDER
This scenario lasts 15 game turns. The Blue player goes first in each turn.

VICTORY CONDITIONS
The Blue player wins by eliminating all enemy units by the end of the game. Failure to do so constitutes a Red victory.

INSPIRATION
Military history has many examples of heroic last stands against insurmountable odds. This scenario owes nothing to any particular historical event, but the heroism displayed in such affairs provides ideal opportunity for any wargames scenario – and it seems appropriate that the final game in this book should be entitled 'Last Stand'. I very much hope that you enjoy playing these games as much as I enjoyed designing them!

FURTHER READING
Two excellent scenarios of this type can be found in the following books:

Grant, Charles Stewart, *Scenarios for Wargames* (Wargames Research Group, 1981) (pp. 36–37).
Grant, C. S. and Asquith, S.A. *Scenarios For All Ages* (CSG Publications, 1996) (pp. 46–47).

Wargame Scenarios 125

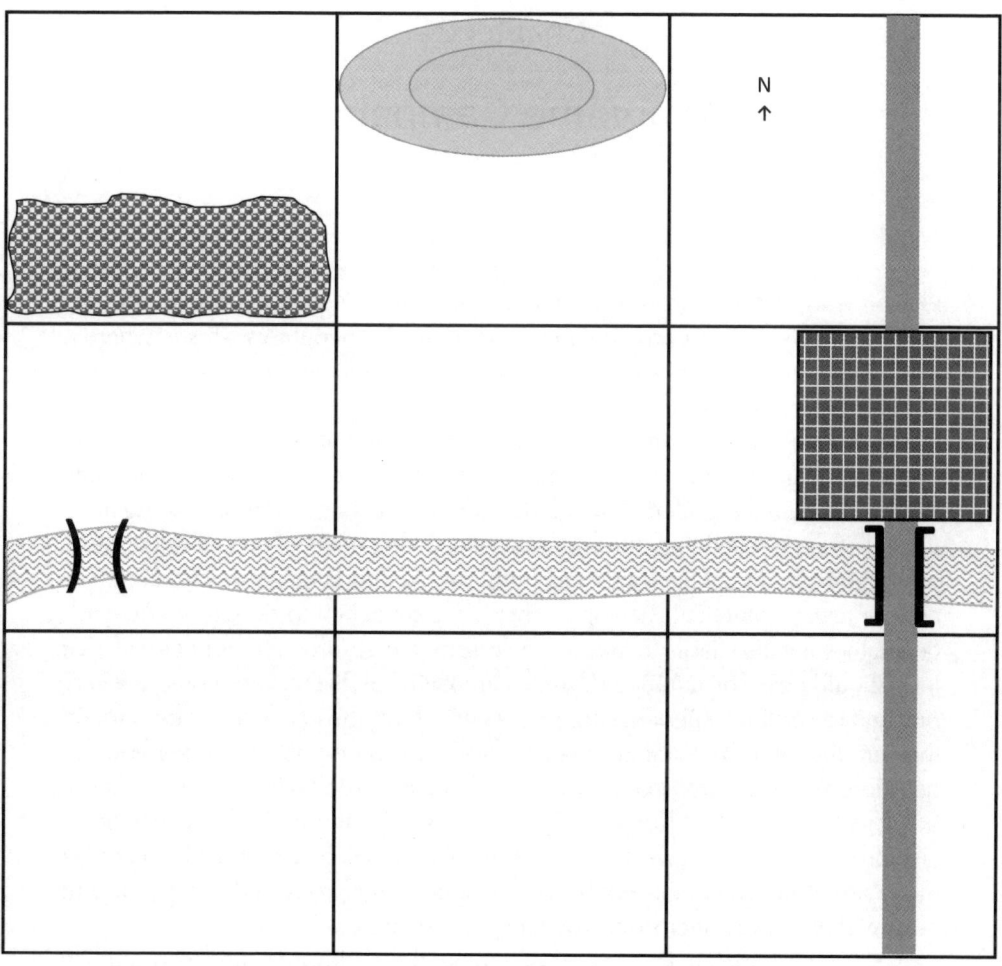

Chapter 21

Wargame Campaigns

Playing wargame scenarios in isolation is always very entertaining, but players will eventually want to provide a context for such encounters. This is where a wargames campaign comes in, when two armies fight over a period for control of their respective territories.

The key to any wargames campaign is feasibility. It is all too tempting to produce detailed maps of the campaign territories (either based on genuine atlases, or lovingly rendered by the designer of the game), and devise detailed rules for forced marching, attrition, sieges, delaying actions, pursuit after battle, promotion of units to elite status, raw militia and reinforcements. Such campaigns are not always practical, having as they do a tendency to drown in excessive detail; they are also liable to take a very long time to resolve. This is not to say that they should never be attempted, since campaigns of this ilk can promote a very rich and rewarding experience for players who have the necessary commitment: they are therefore best suited to people who are so fascinated by the period in question, that they are prepared to devote themselves entirely to it – without being seduced by other historical epochs, or the dictates of the latest wargames fashion. Few players have the necessary endurance or devotion in practice; many are attracted by the notion of a detailed and protracted campaign, but most will drift away after just a few gaming sessions.

A simple campaign is in contrast eminently achievable, and can still be rewarding if done in the right way. The easiest approach is to play an odd number of battles, with the winner of the majority being declared victorious in the campaign. Players could, for example, select three favourite scenarios from the thirty provided in the last chapter, dicing to see who plays Red or Blue at the start of each. This campaign can easily be finished in a long evening, or a half day session at the weekend – the brevity of the games combined with the fact that favourite scenarios are being played, should guarantee interest.

An alternative approach is to generate the relevant scenarios at random. The players could for example decide to play five games, which are determined on the basis of dividing the thirty scenarios into five groups, covering numbers 1–6, 7–12, 13–18, 19–24, and 25–30. One game is played within each grouping of scenarios, with each selected at random. Let us assume for example that a game is being played from scenarios 7–12: this is chosen by rolling a die, with a score of 1 indicating scenario 7, a 2 selecting scenario 8, a 3 choosing scenario 9, and so on. The sides for each turn can be selected randomly, but a viable alternative is for the victor of the previous game to decide whether he or she wants to play Red or Blue in the next scenario. We can for example assume that the campaigners in question start off by playing a game from scenario 1–6: a 3 is rolled on the die, and scenario 3 played accordingly, with the Blue player winning. The next game is from scenarios 7–12; a die roll of 5 indicates scenario 11, and the erstwhile Blue player, having a penchant for attacking, chooses to play Blue again in the Surprise Attack scenario. Allowing the victor to decide which side to choose in the next game owes a great deal to military logic; the winner would undoubtedly have the advantage over the loser in real life, and would therefore be able to seize the initiative in the next encounter. A campaign of five games will last a full day, or two evening sessions.

This style of wargames campaign is very basic, but can be completed quickly and easily. The players' enjoyment of such contests can be greatly enhanced by composing a narrative history of the affair, which adds character to the occasion. Wargamers are encouraged to give their tabletop generals suitable names, be they historical characters such as Napoleon or Wellington, favourite writers or sporting heroes, or even their own surnames suitably adapted. My own name is for example fine in British or French armies, but can easily become 'von Thomas' for a Germanic force or 'Thomassov' for a Russian contingent's commander. The battles can be described in suitably lurid detail, and credibility enhanced by composing an appropriate linking narrative to contextualize them. For example, in the previous case where scenario 3 was followed by scenario 11, the campaign storyline could read something like this:

> 'Having defeated General Prince Turgenev at the battle of River Zola, Count Stendhal's forces pressed home their advantage by attempting to seize Pushkin's Crossroads before Turgenev could organize a response.'

The eventual result will be a full account which, if augmented with artistically rendered maps of the battles in progress, provides a satisfying memento of an enjoyable series of wargames. They can even be used as a spur to campaign design: in reflecting upon the events of the previous game, the players can consider which scenario would be most appropriate to play next, through exercising military logic. This will in the long run encourage the design instincts of players, allowing what started as a very basic campaign to develop into a rich experience as the two belligerent powers (and their assorted generals) acquire a character all of their own. In this way, simplicity leads to the sort of creativity that could be absent from a dauntingly complex campaign – and with creativity comes the originality and enthusiasm that only the best wargames can generate.

Chapter 22

Solo Wargaming

The best wargames will always be achieved when the hobbyist is able to enjoy regular contests with a congenial opponent. This is unfortunately not always possible, in which case the only option is to play solo wargames.

These singular contests are very easy to arrange, at least on the most basic level of playing both sides to the best of one's ability. This style of solo play can produce satisfying and informative games – you can learn a great deal about military history by attempting to execute appropriately realistic tactics on the wargames table, and evaluating their strengths and weaknesses according to the results which occur. True solo wargaming is slightly different, however: this is when the actions of one side are determined by specially devised gaming mechanics (the concept is similar to the 'artificial intelligence' of computer wargames), in an attempt to create the excitement and tension that would otherwise occur when playing a real-life opponent.

Solo wargame systems can be devised for any scenario; it must however be stressed that different approaches have to be devised in each case, to take account of the distinguishing features of various contests. The following examples from my thirty scenarios cover cases where all units begin the game on the table (scenarios 1 and 2); where all arrive on turn 1 (scenario 3); and where one player has only one tactical option (scenarios 7, 13, 14, 15 and 30).

The first two scenarios cover pitched battles, and have the Red player deploy first. In these instances the wargamer takes the role of the Red side, and places the units on the table as he or she desires. The Blue deployment is in contrast determined by the game system, which is intended to strengthen some sectors at the expense of others. It does so by dividing the battlefield into left, centre, and right sections, each of which has a width of 12". A die is then rolled: a score of 1 or 2 indicates that three randomly determined Blue units are deployed in the left sector; 3 or 4 denotes the central portion; and 5 or 6 the right section. A die is then rolled again: a score of 1–3 indicates that two randomly selected units are placed in one of the remaining areas, with a 4–6 denoting deployment in the other. The final unit is positioned in the last sector. This system effectively determines the Blue attack plan, with one area being strong and one weak – it may in so doing create an unwelcome element of surprise for the solo wargamer to confront and hopefully overcome.

Scenario 3 ('Control the River') is slightly different, since the troops begin the game on the table, but all arrive on the first turn. The solo gamer again takes the role of the Red player, marching his or her troops onto the table first. The Blue army's appearance is determined by dividing the table into six sections, each of which is 6" wide, and dicing for each unit to determine precisely where it appears.

Random deployment is a very useful tool that can quite reasonably be used with many scenarios. It is however less necessary in games where one side has only a single tactical approach, be it either tenacious defence (scenarios 7, 14, 15 and 30), or all-out attack (scenario 13). In these cases the plan is fairly obvious, and a good deal less randomization is required. Such caveats notwithstanding, the best solo wargames occur when unpredictability can be added to the encounter, thereby creating an all-important element of tension that might otherwise only be present when playing a live opponent.

One way of doing this is by modifying the reinforcement schedule, so that units do not appear on the designate turn, but that a die is rolled from one turn ahead of schedule: they arrive on a roll of 5–6. Reinforcements may therefore arrive early, they may appear late, or in especially unfortunate cases not at all. The element of unpredictability can be enhanced by having the game system's units appear randomly, so that the active player can never be aware precisely which enemy units may turn up – so that a slow infantry unit may be expected, only for a rapid cavalry contingent to arrive instead, creating potential mayhem in the process.

A very popular and long-standing technique of introducing unpredictability into the solo wargame is through the use of 'Chance Cards'. This very sound approach was devised by great wargaming pioneers such as Donald Featherstone, and takes the form of having a card drawn at the start of each turn for both competing sides: the results can vary from nothing happening, through allowing some great benefits to one's own units, or significant disadvantages. This creates an element of tension, and does in particular realistically invoke the vicissitudes of fortune that can bless or afflict armies in reality. My scenarios can be played using a deck of fifteen cards for each side: one of these is drawn at the start of every Red or Blue turn, their effects being as follows:

Card Numbers	Result
1–5	**No Event.** Nothing happens.
6–7	**Confusion.** 1–3 of one's own units may not move this turn.
8–9	**Ammunition Shortage.** 1–3 of one's own units may not shoot this turn.
10	**Demoralization.** A single unit on one's own side acquires 1–6 hits.
11–12	**Initiative.** A single unit on one's own side may either move twice, move and then shoot, or shoot twice.
13–14	**Rally.** 1–3 of one's own units remove 1–3 hits points.
15	**Enemy Panic.** A single enemy unit acquires 1–6 hits.

Units affected by Chance Cards should always be selected at random, to increase the element of unpredictability. Numbers from 1–3 can be generated by rolling a die and halving the result, rounding up any fractions.

It should be apparent that solo wargaming need not be a poor alternative to traditional social gaming. The element of unpredictability always creates excitement, and the solo wargamer is always free to experiment. He or she can for instance devise new and radical rules and test them thoroughly to assess their viability, without having to worry about the baffled incomprehension of a dissatisfied opponent. The soloist may also want to try playing a new set of wargames rules that may not find favour with anyone else, at whatever level of complexity he or she finds desirable. Solo wargamers can also devise long campaigns involving great depth – this is, as considered in the previous chapter, a significant problem when dealing with the outside commitments of other potential campaign participants. Soloists can also spend a great deal of time and effort exploring and gaming obscure conflicts of very limited appeal to others; I have for example in the past given much consideration to the Warlord era in China (1911–1930) – a period which although fascinating, has attracted decidedly limited interest with most wargamers.

The key to a successful solo wargame is the same as any other: gamers should be creative, remain practical in their aims, and enjoy the rich variety that is guaranteed to result from what will always be a fascinating hobby.

Appendix I

Background Reading

The books listed below are intended to provide further reading for anyone who wants to find out more about either the wargames hobby, or military history. Each section contains a small selection of books, allowing any reader to acquire a substantial but not overwhelming amount of information; my intention has been to choose the best possible works, thereby enabling readers to obtain knowledge and still retain enthusiasm!

Some of the books listed are out of print. They can be obtained from the book dealers listed in Appendix 2, or alternatively from local libraries (often via the inter library loan system).

1. WARGAMES BOOKS

My selection is predicated on the notion that I am choosing books rather than mere rulesets. Many of the latter can be obtained quite easily, but I prefer to list works which have a broader approach, and which cover a great deal more than sets of rules alone – especially those products which are very lengthy, and whose prose is exceptionally legalistic. I do not find such qualities especially appealing; books which include such themes as historical content, wargame scenarios, insights into the design parameters of any rulesets included, and above all simple wargames, are far more attractive than the desiccated pedantry of some rulebooks.

General Titles

Featherstone, Donald F., *War Games* **(Stanley Paul, 1962)**
Before Donald Featherstone wrote this book, the wargames hobby was so small that all its participants were personally acquainted. This magnificent work changed all that. It covers all significant aspects of the hobby: from a description of its essence; through details of how to acquire war games armies and battlefield terrain; and even such fascinating aspects as organizing a wargames campaign. Most importantly, the book included some simple and accessible rules for the

Ancient, Horse and Musket, and Second World War periods – along with a description of wargames battles for each. This book succeeded in popularizing the hobby – it is difficult to see how wargaming could have existed without it.

Grant, Charles, *Wargame Tactics* (Cassell, 1979)
This book does not contain any rulesets, but does include outstanding summaries of military tactics from ancient times to the American Civil War, and most importantly a total of eight wargames battle reports, illustrating how historical tactics are put into effect on the gaming table. The book represents a fine and accessible introduction to military history, as well as its titular subject.

Hyde, Henry, *The Wargaming Compendium* (Pen & Sword, 2013)
This beautifully presented tome provides an ideal introduction to the contemporary hobby. It includes a history of wargaming, guides to figure painting and terrain production, hints on setting up a wargames campaign, and a set of rules for the horse and musket period. The hundreds of outstanding colour photographs help to create a visual feast.

Quarrie, Bruce (ed.), *PSL Guide to Wargaming* (Patrick Stephens, 1980)
This book features some excellent descriptions of all major (and some minor) wargames periods, covering history, tactics, and how to re-enact the relevant epochs in miniature. Each chapter is written by a renowned expert in the relevant field. An interesting set of rules is included, which owes much to the (then) pre-eminent rulebooks published by the Wargames Research Group. They provide for stimulating games, but can require a fair amount of record keeping.

Thomas, Neil, *Wargaming: An Introduction* (Sutton Publishing, 2005)
My own contribution to this genre covers much of the ground included in the book you have just read. The rulesets included are slightly different, but still very simple. I provide very little detail on scenarios, but do cover the history of wargaming, and also enclose army lists for the major forces of each period. The book does in addition examine skirmish wargaming, which includes a ruleset focussed on colonial encounters.

Wesencraft, C.F., *Practical Wargaming* (The Elmfield Press, 1974)
A book that proves how looks can deceive. This is ostensibly just another introductory book with simple rulesets covering all periods up to the end of the nineteenth century, together with advice on how to build up wargames armies and set up a tabletop battlefield. This book is however not only exceptionally well written, but includes design concepts that were decades ahead of their time

– so much so that it was ignored upon publication. This outstanding work has had a massive influence upon my approach to wargame design; the title of my first chapter pays homage to Mr. Wesencraft's magnificent opus.

Ancient Wargaming

Barker, Phil, *Ancient Wargaming* (Patrick Stephens, 1975)
The author is a founder member of the Wargames Research Group, then (and arguably still) the most influential organization in the Ancient wargames field. The book provides an outstanding brief guide to troop types and how they fought, both in history and indeed on the wargames table. It also includes an entertaining history of the development of ancient wargaming, and its generally enthusiastic tone is both infectious and inspirational.

Featherstone, Donald F., *War Games through the Ages Volume 1: 3000BC–1500AD* (Stanley Paul, 1972)
This work focuses upon thirty historical armies, giving details of their tactics and performance, along with how well they are likely to perform on the wargames table. It is an ideal wargamer's guide to its subject.

Grant, Charles, *The Ancient War Game* (A & C Black, 1974)
This is a fine discussion of the principles of wargames rules, and the major armies of the ancient world. It is greatly enlivened by extremely entertaining accounts of the wargame in practice, through the medium of battle reports of some wargame scenarios.

Sabin, Philip, *Lost Battles* (Hambledon Continuum, 2007)
An ambitious and highly successful attempt to assert the credibility of wargaming as a tool for historical understanding. A set of rules is included, but the book's greatest value lies in its discussion of the design principles behind them, and especially in the forensic examination of thirty-five historical battles, particularly with respect to the constituent elements of the contesting armies.

Thomas, Neil, *Ancient and Medieval Wargaming* (Sutton Publishing, 2007)
My own contribution to this genre divides the period into four sections (Biblical, Classical, Dark Ages, and Medieval), providing slightly different rules for each in order to encapsulate and illuminate the relevant differences. I include fifty-three army lists covering all relevant forces, and a wargames battle report from each period (covering the Battles of Kadesh, Issus, Mount Badon, and Agincourt).

Dark Age Wargaming

Donald Featherstone, *War Games through the Ages Volume 1* and my own *Ancient and Medieval Wargaming*, covered in the previous section, provide much valuable information on the Dark Ages. More books are cited below:
Jones, Steve and James Morris, *The Age of Arthur* (Warhammer Historical, 2007)
This work is a sourcebook designed for the *Warhammer Ancient Battles* wargames rules (an absolutely outstanding ruleset, incidentally). It does however contain a great deal of valuable historical information on warfare in Britain from 400 to 800, especially about army composition and tactics.

Mersey, Daniel, *Glutter of Ravens* (Outpost Wargame Services, 1998)
This has all the appearance of a wargaming rules booklet, but is a great deal more than that. It gives some very well informed historical commentary on the Arthurian age (400–700), including wargames army lists and uniform information. A much revised and updated version of the rules has just been published under the new title *Dux Bellorum* (Osprey, 2012).

Patten, Stephen, *Shieldwall* (Warhammer Historical, 2002)
Another sourcebook in the Warhammer historical range, with all the virtues of the work on Arthurian warfare. This volume covers the Viking age from 785 to 1085.

Medieval Wargaming

Books on this period are conspicuous by their absence, with the significant exceptions of Donald Featherstone, *War Games through the Ages Volume 1*, and my *Ancient and Medieval Wargaming*.

Pike and Shot Wargaming

Featherstone, Donald F., *War Games through the Ages Volume 2 1420–1783* (Stanley Paul, 1974)
Another splendid book along the lines of the previously cited Volume One on the ancient period. This work includes a discussion of six major conflicts covered by my Renaissance timeline (1450–1650).

Gush, George and Martin Windrow, *The English Civil War* **(Patrick Stephens, 1978)**
This brief guide to the conflict includes an entertaining wargames battle report, and a particularly fine summary of the organization and tactics of the contending armies.

Wesencraft, C. F., *With Pike and Musket* **(The Elmfield Press, 1975)**
The paucity of works on the Pike and Shot period is more than compensated for by this quite outstanding book. It contains a fine summary of weaponry and tactics (covering English armies from 1547 to 1651), a brilliant set of innovative and simple rules with an erudite design exposition thereof, and guides to the wargames re-enactment of twenty-seven historical battles. This neglected masterpiece is the only wargames book that any reader need consult on the English Civil War; it is unlikely to ever be surpassed.

Horse and Musket Wargaming

Grant, Charles, *The War Game* **(A & C Black, 1971)**
This classic work covers all aspects of wargaming during the eighteenth century. Its best feature can be found in the erudite way that the outstanding wargames rules are explained with reference to historical precedent. The book triumphantly displays that profound scholarship can be displayed without any signs of tedious pedantry.

Griffith, Paddy, *Napoleonic Wargaming for Fun* **(Ward Lock, 1980)**
This masterly work was shamefully neglected upon publication, largely due to its forceful declaration that simplicity was preferable to complexity when wargaming. The book includes seven different sets of rules, covering all levels from minor skirmishes to clashes between armies.

Quarrie, Bruce, *Napoleonic Wargaming* **(Patrick Stephens, 1974)**
This book introduced me to its subject, and provides an admirable summary, covering all aspects of the period from the organization of each nation's army, through a discussion of Napoleonic strategy and tactics, to a useful section on setting up a wargames campaign. The ruleset provided is quite complex, yet concise and admirably playable (although a fair bit of record keeping is required); it is especially enlivened by the use of national characteristics to reflect the strengths and weaknesses of each army.

Thomas, Neil, *Napoleonic Wargaming* (The History Press, 2009)
My own contribution to this subject features substantial historical content, along with commentary on Napoleonic strategy and tactics. The rules are simple, but are preceded by a chapter detailing the principles behind them. The book also includes comprehensive army lists covering all the major forces, and a battle report describing a wargame in action.

Young, Brig. and Lt. Col. J. P. Lawford, *Charge!* (Athena Books, 1986; originally 1967)
It is remarkable that the eighteenth century, though never an especially popular wargames period, should still have inspired two of the greatest classics in wargames literature: Charles Grant, *The War Game*, is one; this is the other. Its enthusiastic tone is especially infectious; it covers two different sets of wargames rules, including substantial discussion of the ideas behind them, and a battle report describing how each game works in practice. The elementary game is especially noteworthy; its rules are just two pages long, but are logically coherent and conceptually perfect – a masterly demonstration that brevity and brilliance can be achieved in wargames rules.

Rifle and Sabre Wargaming

Drewienkiewicz, John and Andrew Brentnall, *Wargaming in History Volume 8* (Ken Trotman, 2013)
This beautifully produced book concentrates upon simulating the opening battles between Austria and Prussia during the Seven Weeks War of 1866. A large selection of maps and detailed orders of battle are included; the book also provides an impressive examination of contemporary tactics, and an annotated bibliography. It represents essential reading for anyone interested in what is a fascinating and unjustly neglected subject.

Featherstone, Donald F; *War Games through the Ages Volume 4: 1861–1945* (Stanley Paul, 1976)
This wargamer's analysis of the period concentrates on the translation of historical tactics to the wargames table. It is an inspirational book for anyone who aspires to design their own rules.

Thomas, Neil, *Wargaming: Nineteenth Century Europe 1815–1878* (Pen & Sword, 2012)
I wrote this book in an attempt to draw attention to a fascinating period that has been scandalously and inexplicably neglected by wargamers. Substantial

historical content is provided, and the simple rules are (as with my Napoleonic book) preceded by a chapter explaining the principles behind them. A large part of this book is however devoted to scenarios, both historical (covering ten battles) and hypothetical – a substantial selection of army lists are included to fight the latter.

Weigle, Bruce, *1870* (Medieval Miscellanea, 2001), *1859* (Medieval Miscellanea, 2006) and *1866* (Medieval Miscellanea, 2010)
These volumes contain workmanlike sets of wargames rules. They are however most valuable for the historical analysis therein; the outstanding sets of scenarios, including some superb maps of historical battlefields; and the massively detailed annotated bibliographies. The first volume covers the Franco-Prussian War; the second examines the Franco-Austrian and Second Schleswig Wars; and the third analyses the Seven Weeks War. The quality of the scholarship and fluency of the prose make all three books essential for the dedicated wargamer.

American Civil War Wargaming

Stevenson, Paul, *Wargaming in History: The American Civil War* (Argus Books, 1990)
This book does not include any rules, but does cover the organization and armament of the contesting sides, along with a particularly valuable discussion of how the armies performed on the battlefield.

Wise, Terence, *American Civil War Wargaming* (Patrick Stephens, 1977)
The author was always a great believer that wargaming should primarily be an enjoyable diversion rather than a governing obsession. This wonderfully entertaining book covers organization and tactics; it also includes a set of simple and accessible rules.

Machine Age Wargaming

There are no books specifically devoted to this period. Donald Featherstone's book cited in the Rifle and Sabre section above does however cover this epoch very well indeed. Readers can easily adopt the rules contained in the Second World War books cited below; a fine dedicated ruleset for the Machine Age is Paul Eaglestone's *A World Aflame* (Osprey, 2012), devoted to the period from 1918–1939.

Second World War Wargaming

Asquith, Stuart, *Wargaming World War Two* **(Argus Books, 1989)**
Another thoroughly entertaining book, whose author believes in the virtues of simple wargames, this work includes much useful information on the organization and weaponry of conflict's various armies, but is especially valuable in its provision of simple rules and basic scenarios – the author does not confine himself to land operations, but includes games covering naval and aerial warfare too.

Grant, Charles, *Battle! Practical Wargaming* **(Model & Allied Publications, 1970)**
This book includes an excellent set of accessible rules, along with a historically well informed discussion of the principles behind them. It includes some entertaining battle reports, which show how the wargame works in practice.

Lyall, Gavin and Bernard Lyall, *Operation Warboard* **(A & C Black, 1976)**
This is the first wargamers book I read, and its virtues encouraged me to pursue the hobby a good deal further than I would ever have expected! The book includes details on how to create wargame armies, highly atmospheric wargames battle reports, and an entertaining set of rules with an illuminating exposition thereof. Gavin Lyall was a journalist and thriller writer by profession; this explains both the literary flair and clarity of his prose style.

Quarrie, Bruce, *World War 2 Wargaming* **(Patrick Stephens, 1976)**
Many wargamers who examine the Second World War have a love of complex rules and technical minutiae. This enormously influential book is very much for them; my tastes may differ widely, but it would be churlish not to acknowledge what was a seminal work of its kind. An entire generation of wargamers grew up using this book, and play games of a similar outlook to this day.

Scenarios

Drewienkiewicz, John and Adam Poole, *Wargaming in History Volumes 3 and 6* **(Ken Trotman, 2001 and 2012)**
These beautifully produced books present wargame scenarios and battle reports covering several engagements from the American Civil War battles of Gettysburg (volume 3) and First Bull Run (volume 6). The maps, orders of battle, and annotated bibliographies are particularly impressive.

Featherstone, Donald F., *Battle Notes for Wargamers* (David & Charles, 1973)
An outstanding selection of fifteen historical battles presented as wargame scenarios written with the verve and clarity that are characteristic of the father of modern wargaming. Two of Donald Featherstone's other books, *Wargaming: Ancient and Medieval* (David & Charles, 1975) and *Wargaming: Pike and Shot* (David & Charles, 1977), do the same thing for their respective periods.

Grant, Charles Stewart, *Scenarios for Wargames* (Wargames Research Group, 1981)
This book of fifty-two wargame scenarios is a model of its kind. My massive debt to its influence is apparent from the reading list appended to many of the scenarios that you have just consulted.

Grant, C. S. and S. A. Asquith, *Scenarios For All Ages* (CSG Publications, 1996)
Another fine selection of scenarios, along the lines of the work just cited.

Grant, Charles S. et al, *Wargaming in History Volumes 1, 2, 4, 5, and 7* (Ken Trotman, 2009–2012)
These magnificent books represent the zenith of modern wargames writing. They are devoted to presenting wargames re-enactments of historical battles from the Seven Years War (volumes 1, 4, and 5), the War of the Austrian Succession (volume 2), and the Peninsular War (volume 7). They are notable not only for the clarity and viability of the scenarios, but also for the beautiful photographs of the games; artwork of assorted troop types by the acclaimed artist Bob Marrion; and concise yet profound discussion of wargames rules writing and design.

Campaigns

Bath, Tony, *Setting up a Wargames Campaign* (Wargames Research Group, 1978)
The author's Hyborian campaign, set on an imaginary continent of his own devising, is the most famous contest in the history of Ancient wargaming. This book contains all the rules for the campaign in question, and has many very useful tips on such aspects as pursuit after battle, sieges, delaying actions, supply and recruitment. Although focussing on the Ancient period, the book does contain suggestions on adopting the rules for other epochs. Very much a book designed for players who wish to absorb themselves in a campaign of long duration.

Featherstone, Donald F, *War Game Campaigns* **(Stanley Paul, 1970)**
This is the classic account of its subject. It covers all varieties of campaign from small to large, and considers all historical periods. It is moreover written with all the author's customary enthusiasm and verve, and is a truly inspirational read.

Grant, Charles S; *Raid on St. Michel* **(Partizan, 2008);** *The Annexation of Chiraz* **(Parizan, 2009); and** *The Wolfenbüttel War* **(Partizan, 2012)**
These are very entertaining books (the first two of which were written in conjunction with Phil Olley), cover narrative campaigns based upon a hypothetical eighteenth century setting. They provide a model of how to create a simple wargames campaign.

Solo Wargaming

Asquith, Stuart, *The Partizan Press Guide to Solo Wargaming* **(Partizan, 2006)**
This book provides a wonderfully concise introduction to all aspects of its subject, with the author's customarily lively and accessible style making it a pleasure to read.

Featherstone, Donald, *Solo-Wargaming* **(Kaye and Ward, 1973)**
Another classic account from the father of modern wargaming, this book covers absolutely everything anyone needs to know, and is full of stimulating ideas. As with Stuart Asquith's book just cited, it is an absolute joy to read this seminal work.

Painting, Terrain and Uniforms

Dallimore, Kevin, *Foundry Miniatures Painting and Modelling Guide* **(Foundry, 2006)**
This book provides an excellent guide for anyone who wishes to paint wargames figures to a high standard.

Games Workshop, *How To Make Wargames Terrain* **(Games Workshop, 2005)**
An excellent and highly accessible guide for all readers, not simply the fantasy and science fiction wargamers that Games Workshop generally focus upon.

Kannik, Preben, *Military Uniforms in Colour* **(Blandford, 1968)**
This incredibly useful book provides over five hundred colour illustrations of troops from the Horse and Musket period to the Second World War.

Background Reading 141

Osprey Publishing, Various Titles
This publisher produces a vast range of illustrated books featuring the uniforms of troops from all ages. Readers are strongly advised to consult the relevant works from periods of particular interest.

2. MILITARY HISTORY

A comprehensive list of military history titles would be so vast as to be impossible for anyone to read. I have largely confined myself to titles dealing with how armies fought, thereby allowing readers to appreciate how accurate any wargame really is – and hopefully, to encourage all of you to start designing, writing and playing games with your own rules.

General Works

Dupuy, R. Ernest and Trevor N. Dupuy, *The Collins Encyclopaedia of Military History* (BCA, 2007; originally 1993)
This is more of a chronological reference guide to conflict rather than a traditional encyclopaedia (it lacks the customary A – Z entries). It does give fine provision of the basic facts relating to all major conflicts, and its summaries of significant military developments are always very useful.

Howard, Michael, *War in European History* (Oxford, 1976)
A brief, well written and very scholarly account of strategic developments from Medieval times to the present day. The quality of the analysis is extraordinarily high.

Keegan, John, *The Face of Battle* (Penguin, 1976)
This justifiably renowned book deals with the experience of battle at Agincourt, Waterloo and the Somme, and represents an outstandingly valuable source for any wargamer. The frequently overlooked first chapter, dealing with the utility and limitations of military history as an intellectual discipline, is also well worth consulting.

Ancient Warfare

Anglim, Simon et al, *Fighting Techniques of the Ancient World* **(Greenhill, 2002)**
This is an exceptionally useful introduction on how the principal forces of the period were equipped, and how they fought on the battlefield. Its five sections cover infantry warfare, cavalry warfare, command and control, sieges, and naval engagements.

Barker, Phil, *The Armies and Enemies of Imperial Rome* **(Wargames Research Group, 1981)**
This quite outstanding title has achieved almost legendary status among veteran ancient wargamers. It gives a brilliant analysis of the organization, tactics, dress, and weaponry of all the many armies covered – the quality of which is greatly enhanced by Ian Heath's black and white illustrations of the many troop types involved.

Connolly, Peter, *Greece and Rome at War* **(Greenhill, 1998)**
This is a very comprehensive examination of its subject, which is especially strong on the organization and weaponry of the armies covered. The many colour illustrations of the assorted troop types provide an excellent guide for wargames figure painting.

Hanson, Victor Davis, *The Western Way of War* **(Hodder & Stoughton, 1989)**
A forensic analysis of the Greek hoplite's place in society, along with his role and effectiveness on the battlefield. Hanson's arguments are exceptionally wide ranging, and have proved to be extremely influential.

Pietrykowski, Joseph, *Great Battles of the Hellenistic World* **(Pen & Sword, 2009)**
This book covers all the major battles of Alexander the Great and the Greek world in general, up until the Roman conquest. Each battle's campaign, topography, and army composition are covered; the description of the respective engagements is notable for its great verve, which gives the reader an impression that he or she is actually present at the battle – the process is greatly assisted by the clear and comprehensive plans of each engagement as it progresses.

Dark Age Warfare

Aitchison, Nick, *The Picts and the Scots at War* (Sutton, 2003)
This book examines every aspect of its hitherto neglected subject, and provides a rigorous and scholarly treatment of its themes.

Heath, Ian, *Armies of the Dark Ages 600–1066* (Wargames Research Group, 1980)
This book provides an excellent introduction to the organization, tactics, dress and weapons of all the armies of the period. The author has provided many line drawings of the assorted troop types, and this book is an absolutely essential reference source for all wargamers.

Hill, Paul, *The Anglo-Saxons at War* (Pen and Sword, 2012)
This comprehensive and highly accessible book is an invaluable source for anyone interested in Dark Age warfare in Britain.

Oman, Sir Charles, *A History of the Art of War in the Middle Ages Volume One: 378–1278 AD* (Greenhill, 1991; originally 1924)
This is a classic account of its subject. Readers could quite easily find out all they need to know from this book alone; it covers all seminal developments in a manner both impressively scholarly and exceptionally readable. It also represents an excellent source for the following Medieval period.

Peers, Chris, *Offa and the Mercian Wars* (Pen and Sword, 2012)
This book covers the wars of the Mercian Saxon kingdom from 600 to 875. It therefore represents an ideal precursor to Paul Hill's book on the later period from 800. The author is a wargames writer of some eminence; he has a penchant for designing rulesets that are simple yet highly original. These qualities mean that his historical works are of great interest and value for all wargamers.

Medieval Warfare

Bennett, Matthew et al, *Fighting Techniques of the Medieval World AD 500– AD 1500* (Spellmount, 2005)
A book which focuses on battlefield activity will inevitably be of interest to all wargamers. This attractively illustrated volume covers all aspects of warfare, with the sections on cavalry warfare and leadership being the most valuable.

Heath, Ian, *Armies of Feudal Europe 1066–1300* (Wargames Research Group, 1978)
A work along exactly the same lines as the author's previously cited book on Dark Age armies, and one that is equally essential for any wargamer.

Keen, Maurice (ed), *Medieval Warfare* (Oxford, 1999)
A team of eminent scholars combined to produce this academic introduction to its subject. The book is divided into two sections, the first of which provides a useful chronological survey; the second part is devoted to chapters on specific themes, of which Andrew Ayton's account of mounted knightly warfare is especially informative.

Pike and Shot Warfare

Eltis, David, *The Military Revolution in Sixteenth-Century Europe* (I. B. Tauris, 1995)
The question over the nature, extent and timing of a European military revolution in the Renaissance age, has given rise to much academic discussion ever since Michael Roberts gave a lecture on the subject in 1956. This book provides all the rigour and scholarship one would expect from an academic analysis; it is however much better written than most, and includes valuable information on battlefield tactics.

Gush, George, *Renaissance Armies 1480–1650* (Patrick Stephens, 1975)
This book is very similar to Phil Barker's *The Armies and Enemies of Imperial Rome*, in that it describes the organization, equipment and tactics of all armies in a concise, yet comprehensive and erudite manner – again assisted by Ian Heath's illustrations. This book has justifiably acquired legendary status amongst wargamers of the Pike and Shot epoch.

Jörgensen, Christer et al, *Fighting Techniques of the Early Modern World* (Spellmount, 2005)
This follow-up volume to the previously cited work on *Fighting Techniques of the Ancient World* covers the same themes, and does so with similar effectiveness.

Oman, Sir Charles, *A History of the Art of War in the Sixteenth Century* (Greenhill, 1987; originally 1937)
This is one of the greatest military history books ever written. It covers the equipment, tactics and effectiveness of all significant European armies, along

with a description and examination of every major battle. Every subsequent book covering this period owes a massive and incalculable debt to Sir Charles' work.

Seymour, William, *Battles in Britain 1066–1746* (Wordsworth, 1977)
A fine examination of the relevant engagements, greatly assisted by the very clear battle plans. This is a particularly useful source for the engagements of the English Civil War.

Horse and Musket Warfare

Chandler, David G., *The Campaigns of Napoleon* (Weidenfeld and Nicolson, 1966)
Many books have been written on the Napoleonic Wars; this is the only one that you really must read. It provides a masterly analysis of every campaign and battle that Napoleon Bonaparte ever fought; the breadth of the theme is matched by the depth of the scholarship and the quality of the prose.

Chandler, David, *The Art of Warfare in the age of Marlborough* (Spellmount, 1990; originally 1976)
This book gives a detailed account of infantry, cavalry, and artillery at the turn of the eighteenth century. Its account of the battlefield performance of each arm is especially illuminating.

Duffy, Christopher, *The Military Experience in the Age of Reason* (Wordsworth, 1998; originally 1987)
This work paints an evocative portrait of how armies were recruited, their governing ethos, and most importantly how well they performed. The chapter describing and examining the unfolding of a typical battle is outstanding, and is especially useful for wargame designers.

Griffith, Paddy, *Forward into Battle* (Antony Bird, 1981)
This book covers battlefield tactics from the age of Napoleon to Vietnam. The late author always had a reputation for being especially controversial, and certainly excelled himself on this occasion. He was however also a thought provoking writer whose analysis and prose were invariably brilliant. This book argues that military historians have always overrated the importance of firepower on the battlefield, and that the threat of hand-to-hand combat was what really precipitated the rout of armies.

Nosworthy, Brent, *Battle Tactics of Napoleon and his Enemies* **(Constable, 1995)**
An astonishingly detailed theoretical and practical examination of its subject. The author's background as a board wargame designer makes his work especially valuable.

Rifle and Sabre Warfare

Barry, Quintin, *The Road to Königgrätz* **(Helion, 2010)**
This minutely detailed and impressively scholarly account covers all aspects of the Seven Weeks War, including the campaigns in Bohemia, Western Germany and Italy. It also examines the Prussian attack on Denmark during the Second Schleswig War of 1864.

Craig, Gordon A, *The Road to Königgrätz* **(Weiderfeld and Nicolson, 1964)**
A classic account of the campaign in Bohemia of 1866. It is both scholarly and exceptionally readable, being guaranteed to provide both enlightenment and entertainment.

Drury, Ian, *The Russo-Turkish War 1877* **(Osprey,1994)**
The Osprey Men at Arms titles are renowned for their provision of colour uniform illustrations. This work does a great deal more than that. It includes a brief account of the war, detail on the organization of the respective armies, and information on their weaponry.

Glover, Michael, *Warfare from Waterloo to Mons* **(Book Club Associates, 1980)**
An outstandingly written introduction to warfare on land and sea, covering all the salient points extremely well.

Howard, Michael, *The Franco-Prussian War* **(Routledge, 2002; originally 1961)**
This groundbreaking work has won classic status. Its erudition is remarkable; its judgements invariably considered; and its literary quality immense. Any student of the France-Prussian War should read this masterpiece.

The American Civil War

Griffith, Paddy, *Battle Tactics of the American Civil War* **(Crowood, 1989)**
A predictably controversial and brilliant book from this outstanding thinker, Griffith's thesis is that the war's engagements degenerated into prolonged

firefights not as a result of the potency of new weaponry, but because of the indiscipline of the hastily recruited combatants.

Keegan, John, *The American Civil War* (Hutchinson, 2009)
This solid analytical examination of the conflict provides ideal background material.

Nosworthy, Brent, *The Bloody Crucible of Courage* (Constable, 2005)
Another fine theoretical and practical examination of battlefield tactics, along the lines of the author's book on the Napoleonic Wars.

Perello, Christopher, *The Quest for Annihilation* (Strategy & Tactics, 2009)
The series of case studies in this book provide a fine picture of how the armies performed on the battlefield. The author is a board wargame designer whose games are frequently simple yet penetrating – the tendency to get to the point of the issue with rapid clarity is equally apparent in this book.

Machine Age Warfare

Belfield, Eversley, *The Boer War* (Lee Cooper, 1975)
A very clear and comprehensive, yet concise, survey of the war which saw the advent of magazine rifles and heavy artillery. The maps of the various battles are especially good, allowing for easy re-creation as wargames.

Griffith, Paddy, *Battle Tactics of the Western Front* (Yale, 1994)
This description of the revolutionary nature of British tactics during the latter period of the First World War proved to be a revelation upon publication. It demonstrates just how effective the British army became, and how the First World War, far from being a calamitous display of military incompetence, instead represented one of the greatest historical triumphs of British military achievement. This may seem a singular view; the author's erudition and eloquence does however present a very convincing case.

Hughes, Ben, *They Shall Not Pass!* (Osprey, 2011)
This book provides a detailed account of the heroic struggle of British anti-fascist volunteers against General Franco's forces during the Spanish Civil War, at the Battle of Jarama. The author's account comes alive thanks to a good selection of battlefield maps, and a wide range of primary source material.

Stevens, Philip, *The Great War Explained* **(Pen and Sword, 2012)**
A book that lives up to its title, providing as it does a clear and accessible introduction to the conflict – albeit that it only covers the Western Front and Gallipoli. Among its many virtues are its ability to record both calamities and achievements, whilst avoiding intemperate critiques of the one or excessive praise of the other. The section on the Somme is especially useful, as are the appendices covering the leading personalities, and the weapons of the conflict.

Swinton, E. D., *The Defence of Duffer's Drift* **(Leo Cooper 1990; originally 1907)**
One of the most original military history books ever written. The author (who later became famous for being a leading figure in the invention of the tank) examines how an imaginary river crossing could be defended against an attacking Boer force. The revelation that Victorian tactics proved disastrous in the twentieth century environment, and that an entirely counter-intuitive approach was essential, proved to be exceptionally influential. The book is also written in a marvellously entertaining way, and is an absolute classic.

Second World War

Bull, Stephen, *Second World War Infantry Tactics* **(Pen & Sword, 2012)**
A comprehensive examination of battlefield activity in Western Europe, covering the British, German, and American armies. The final appendix, which is a copy of a British military examination of German infantry tactics (published in 1941), is absolutely fascinating, and invaluable for wargame designers.

Ellis, John, *World War II: The Sharp End* **(Windrow & Greene, 1990)**
A comprehensive and illuminating description of the combat experience of British and American soldiers. This book creates a vivid impression of the privations all combatants had to experience.

Fuller, J. F. C., *The Second World War* **(Da Capo, 1993; originally 1948)**
The author was a great pioneer of armoured warfare theory and practice during his army career; the profundity of his thought and the vehemence of its expression succeeded in arousing the ire of his superiors to such an extent, as to precipitate his departure from the British army. Fuller's many works on military history display all the traits that got him into serious trouble with his rather hidebound superior officers; this book provides an especially acute examination of generalship, and is all the more remarkable for displaying such profundity in a work published just three years after the end of the conflict.

Macksey, Kenneth, *Tank Tactics 1939–1945* (Almark, 1976)
A concise examination of its subject, greatly illuminated by the very clear diagrams provided. This is an especially valuable book for any budding wargame designer.

Marshall, S. L. A., *Men against Fire* (Oklahoma, 2000; originally 1947)
The author was a serving American army officer, who spent much of the Second World War interviewing soldiers and finding out about their combat experience. His astonishing discovery that only a quarter of front line troops ever fired their weapons, led to this book. It provides an outstanding examination of how and why combat units performed as they did on the battlefield – and how the essential aggressive intent can be inculcated into serving soldiers. Marshall's book is essential reading for anyone who wants to find out what really happened on the battlefield.

Appendix II

Useful Addresses

The following addresses, telephone numbers and website details will give any wargamer an idea of where to start collecting. Readers should bear in mind that addresses and other contact details can change; those with internet access are strongly advised to check updated information via the relevant website. Also, note that some companies are unable to receive personal callers – so if you do plan to make a visit, always check whether this is allowed before you arrive!

MAGAZINES

These are essential sources of information, advice and inspiration for any wargamer. All three of the journals listed below are available in most large newsagents in the United Kingdom. Subscription enquiries should be made to the following addresses:

Miniature Wargames, Subscriptions: West Street, Bourne, Lincolnshire, PE10 9PH
Tel: 01778 392494
Web: www.miniwargames.com

Wargames Illustrated, Unit 4c, Tissington Close, Beeston, Nottingham, NG9 6QG
Tel: 0115 704 3250
Web: www.wargamesillustrated.net

Wargames: Soldiers & Strategy, PO Box 4082, 7200 BB Zutphen, The Netherlands
Tel: +44-20-88168281
Web: www.wssmagazine.com

MILITARY BOOK SUPPLIERS

Caliver Books, 100 Baker Road, Newthorpe, Nottingham, NG16 2DP.
Tel: 0115 938 2111
Web: www.caliverbooks.com

A vast range of new books and wargames rules can be obtained from this company.

David Lanchester's Military Books, 6 Pinfold Way, Sherburn-in-Elmet, North Yorkshire, LS25 6LF
Tel: 01977 684234
Web: www.davidlanchestermilitarybooks.co.uk

This dealer is a splendid source for quality second-hand books and new publications at bargain prices.

Ken Trotman Ltd., PO Box 505, Huntingdon, PE29 2XW.
Tel: 01480 454292
Web: www.kentrotman.com

A fine range of new military history books can be obtained, but the firm are most renowned for their vast second-hand stock; and also their own publishing arm, which produces a selection of primary source accounts.

Monarch Military Books and Miniatures, Unit 5a, Cuthbert Court, Off Norwich Road, Dereham, Norfolk, NR19 1BX.
Tel: 01362 691435
Web: www.monarchmilitarybooks.com

This trader supplies wargames rules and new books – discounts are frequently offered on the latter.

Paul Meekins Military and History Books, Valentines, Long Marston, Stratford upon Avon, Warwickshire, CV37 8RG
Tel: 01789 722434
Web: www.paulmeekins.co.uk

This dealer can be relied upon to supply many outstanding books that are no longer in print, and is an especially good source for tracking down the wargaming classics. He also sells all the major new releases.

MODEL FIGURE MANUFACTURERS

There are many manufacturers of high quality wargames figures currently trading. Readers are strongly advised to follow up any advertisements in wargames journals, or better still inspect any relevant trader's wares at a local wargames show. The companies listed below do however provide a good starting point for any budding wargamer:

Baccus 6mm Ltd, Unit C, Graham House, Bardwell Road, Sheffield, 53 8AS.
Tel: 0114 272 4491
Web: www.baccus6mm.com

As the name suggests, Baccus only make 6mm figures, the quality of which is outstanding. They also produce an innovative set of wargames rules for the Franco-Prussian War.

Essex Miniatures, Unit 1, Shannon Centre, Shannon Square, Thames Estuary Industrial Estate, Canvey Island, Essex, SS8 0PE.
Tel: 01268 682309
Web: www.essexminiatures.co.uk.

Essex supply a large range of 15mm and 25mm figures from all periods, together with a variety of wargaming accessories, including terrain. They also have an in-house figure painting service.

Foundry Miniatures Ltd., 24–34 St. Marks Street, Nottingham, NG3 1DE.
Tel: 0115 841 3000
Web: www.wargamesfoundry.com.

Probably the largest figure company in the business, Foundry only produce 28mm miniatures. They do however have a very wide range, and the quality is outstanding. Unfortunately, they are more expensive than other manufacturers.

Games Workshop, Willow Road, Lenton, Nottingham, NG7 2WS
Tel: 0115 91 40000
Web: www.games-workshop.co.uk

This phenomenally successful company has many shops in the United Kingdom and worldwide. It deals exclusively in 28mm fantasy and science fiction figures, but some of these can be used in historical wargames (most notably the 'Empire' range for the *Warhammer* fantasy game, which can be used in Pike and Shot encounters). Readers are strongly advised to visit their nearest store and look at what is available – those who wish to play fantasy or science fiction wargames can benefit from purchasing the *Warhammer* and *Warhammer 40000* rulebooks, and from reading Games Workshop's monthly magazine *White Dwarf*. Potential customers should however be aware that this company's products may be of outstanding quality, but are extremely expensive.

Irregular Miniatures, 41 Lesley Avenue, York, YO10 4JR.
Tel: 01904 671101.
Web: www.irregularminiatures.co.uk.

This company produces an incredibly wide range of competitively priced figures in 2, 6, 10, 15, 20, 28, 42 and 54 mm sizes.

Old Glory Corporation, Institute House, New Kyo, Stanley, County Durham, DH9 7TJ.
Tel: 01207 283332
Web: www.oldgloryuk.com.

Old Glory do produce a wide range of 15mm and 25mm figures, but are most famous for manufacturing the largest selection of 10mm miniatures currently available. They have also introduced a selection of 40mm figures.

Spencer Smith Miniatures, The Old Rectory, Wortham, Diss, Norfolk, IP22 1SL.
Tel: 01379 650021
Web: www.spencersmithminiatures.co.uk

This venerable company was one of the great pioneers of wargaming, and its very cheap if slightly basic 30mm horse and musket figures are still available today. Spencer Smith also distribute the famous Willie and Tradition 30mm ranges.

Warlord Games, T04/10 Technology Wing, The Howitt Building, Lenton Boulevard, Nottingham, NG7 2BY
Tel: 0115 978 4202
Web: www.warlordgames.com

This new company produces a growing range of 28mm miniatures, both in metal and hard plastic. They are also notable for publishing four very popular wargames rulesets: *Hail Caesar* (ancient period); *Pike & Shotte* (self-defined); *Black Powder* (Horse and Musket); and *Bolt Action* (Second World War).

Warrior Miniatures, 14 Tiverton Avenue, Glasgow, G32 9NX
Tel: 0141 778 3426
Web: www.warriorminiatures.com

This firm produces a wide range of 15mm and 25mm figures, all of which are extremely cheap.

PLASTIC FIGURE STOCKISTS

Readers are strongly advised to visit their local toy or model shop in order to access plastic figures. The following dealers are however guaranteed to fill in any gaps: both offer every new figure available in 1:76, 1:72 and 1:32 scale soft plastic figures, as well as 28mm hard plastic miniatures; Harfields do in addition sell a large selection of second-hand stock.

F & S Scale Models, 227 Droylsden Road, Audenshaw, Manchester, M34 5ZT
Tel: 0161 370 3235
Web: www.fandsscalemodels.co.uk

Harfields Military Figure Specialists, 32 St. Winifreds Road, Biggin Hill, Westerham, Kent, TN16 3HP.
Tel: 01959 576269
Web: www.harfields.com

PAINTED FIGURE STOCKISTS

Hinds Figures Ltd., 99 Birchover Way, Allestree, Derby, DE22 2QH
Tel: 0133 255 9025
Web: www.hindsfiguresltd.com

This dealer has a vast stock of painted second-hand figures in all sizes, but especially in 15mm and 25mm.

WARGAMES TERRAIN MANUFACTURERS

Magister Militum, Unit 4, The Business Centre, Morgans Vale Road, Redlynch, Salisbury, SP5 2HA.
Tel: 01725 510110
Web: www.magistermilitum.com

A vast selection of scenery is available, both pre-painted and unpainted.

Index

Airfix, 3–4
American Civil War, 46–7
Anglo-Saxons, 16–17
Asquith, Stuart, 80, 112, 116
Auberoche, Battle of, 108

Bismarck, Otto von, 51
Bladensburg, Battle of, 108
Blücher, Marshal, 82

Ceresole, Battle of, 66
Chance Cards, 129–30
Charlemagne, Emperor, 17, 21
Chivalry, 21
Christianity, 15, 21, 27, 33
Classical age, 7–8
Clermont, Count of, 102
Crossbows, 22

Dark Ages, 15–19, 133–4, 143–4
Dunnigan, James F., 96

El Molino del Rey, Battle of, 120
Entrenchments, 53

Featherstone, Donald, 104, 108
Ferdinand of Brunswick, 102
Feudalism, 15, 21
Fontenoy, Battle of, 94
Franks, 15, 17
Frederick the Great, 78

Games Workshop, 5–6
Gitschin, Battle of, 84
Grant, Charles, 95, 106
Grant, Charles Murray, 108
Grant, Charles Stewart, 74, 78, 88, 90, 95, 108, 110
Grant, General, 46
Greece, 7–8

Horse and Musket period, 35–5

Industrialization, 41, 46, 51

Kernstown, Battle of, 114
Knights, 21–2
Krefeld, Battle of, 102

Langensalza, Battle of, 100
Leuthen, Battle of, 78
Lobositz, Battle of, 106
Lundy's Lane, Battle of, 80

McClellan, General, 46
Machine Age, 51–2
Machine guns, 51–2, 57
Medina de Rio Seco, Battle of, 122
Mexican-American War, 120
Middle Ages, 21–2

Napoleon, Emperor, 127
Ney, Marshal, 86

Panzer Blitz, 96
Panzer Leader, 92
Persian Empire, 7
Pocock, John, 100

Quarrie, Bruce, 4
Quatre Bras, Battle of, 86

Renaissance period, 27–8
Rifle and Sabre period, 41–2
Rifled firearms, 41, 47, 57

Salamanca, Battle of, 76
Shiloh, Battle of, 118
Somme, Battle of the, 53
Stalin, Joseph, 15
Stokes, Julian, 96
Strategy and Tactics, 114

Tanks, 52–3, 59
Thames River, Battle of, 112
Thirty Years War, 34

Vikings, 16–17

Wargamer's Newsletter, 104
War of 1812, 112, 118
Waterloo, Battle of, 82, 86
Wavre, Battle of, 82
Wellington, Duke of, 76, 82, 86, 127
World War II, 57

Yellow Ford, Battle of the, 110

Notes

Dear Reader,

We hope you have enjoyed this book, but why not share your views on social media? You can also follow our pages to see more about our other products: facebook.com/penandswordbooks or follow us on X @penswordbooks

You can also view our products at www.pen-and-sword.co.uk (UK and ROW) or www.penandswordbooks.com (North America).

To keep up to date with our latest releases and online catalogues, please sign up to our newsletter at: www.pen-and-sword.co.uk/newsletter

If you would like a printed catalogue with our latest books, then please email: enquiries@pen-and-sword.co.uk or telephone: 01226 734555 (UK and ROW) or email: uspen-and-sword@casematepublishers.com or telephone: (610) 853-9131 (North America).

We respect your privacy and we will only use personal information to send you information about our products.

Thank you!